Automotive
Electrical
HANDBOOK

Jim Horner

HPBooks
a division of
PRICE STERN SLOAN
Los Angeles

HPBooks

Cover photo: Bill Keller

Published by HPBooks
a division of Price Stern Sloan, Inc.
11150 Olympic Boulevard
Los Angeles, California 90064
© 1986 Price Stern Sloan, Inc.
Printed in the U.S.A.

16 15 14 13 12 11 10 9

Thanks

Thanks to the many automotive electrical suppliers and companies who provided photographs, drawings and technical information on their products. Especially helpful were Bob Schlanger of Import Parts International, Dale Penn of Dale's Historic Vehicle Repair, Charley V. Frank of California Terminal Products and John Covan of Longacre Automotive Racing Products.

Special thanks to our technical consultant Frank Parker for providing invaluable information and advice on wiring race cars, and for reviewing the manuscript for accuracy and completeness.

Contents

Notice: The information contained in this book is true and complete to the best of our knowledge. All recommendations on parts and procedures are made without any guarantees on the part of the author or Price Stern Sloan. Because the quality of parts, procedures and methods is beyond our control, author and publisher disclaim all liability incurred in connection with the use of this information.

Introduction

Author Jim Horner beginning wiring job on Ron Fournier's street rod. Fournier is a professional race-car fabricator and author of HP's *Metal Fabrication Handbook.* Whether you're wiring a race car, kit car or other vehicle from scratch, rewiring a collector car, or just wiring in a few accessories on the family car, this book shows you how.

What is this book all about? Well, let's say you're planning to assemble a kit car, also called *component* car or *replicar.* You select the kit of your dreams. Then you pick up a "donor" car to provide the running gear. You lay out your tools and round up the other paraphernalia necessary to turn that pile of fiberglass and hardware into a real, driveable car.

If you're only vaguely familiar with engine rebuilding, brake rebuilding, transmission rebuilding, and suspension assembly and alignment, you'll probably visit a bookstore for the appropriate manuals—maybe even one or two of HPBooks' automotive guides. So far, so good. If you don't know how to do something, you look it up and there's the procedure in black and white.

Work progresses nicely—until you realize your dream car is nearly done, but there's one critical thing missing. Until just this minute, you've completely overlooked the electrical system! With no fire in the wire, there's no heat in the street.

Panic! You contact the kit manufacturer for help. With a sinking feeling, you hear them tell you that if *they* knew how to install an electrical system, they would have included the information in their assembly manual.

You find that your local kit-car club isn't much help either. Their advice—the universal solution—is to *jury-rig* the wiring harness from a donor car and hope for the best. This is hardly worthy of the craftsmanship you put into the show-quality paint and upholstery, the zoomy wheels and tires, and the super-trick engine work. All this good stuff, supported by an electrical system that looks like an explosion in a wire factory? No way!

Do I need to tell you that help is at hand? In fact, *in your hand* right now?

"But," you may protest, "I don't know the first thing about electricity. In fact, I'm almost afraid to change a light bulb for fear of getting shocked!"

This book has been prepared specifically for you. It explains all you need to

know to wire your car from scratch. You'll learn how to select and install electrical components, design circuits, construct and install wiring harnesses, and test the systems.

To enable you to accomplish these tasks, this book also explains electrical theory, wiring diagrams and test procedures. It also includes a comprehensive list of suppliers for just about everything you'll need for doing the job.

What if you aren't wiring a car from scratch? Maybe you're restoring an antique or collector car. Or, perhaps you want to replace or repair a tired wiring harness or specific components in your old family car. If so, you'll find a special chapter on rewiring that tells you how to restore the electrical system to like-new condition. This book also includes several chapters pertaining to selected vehicles—wiring a race car, trailer wiring, and converting a VW 6-volt system to a 12-volt system.

One final note: Because this book

Alternator is source of current for all electrical components while engine is running. It also recharges battery, which is the other electrical source in a car. This 90-amp job could be used as power source for bumper full of driving lamps like those at right. Photo by Tom Monroe.

Lamps and electrical motors are the two major types of electrical loads found on cars. High-power driving lamps like these use a lot of juice, so battery and alternator (sources) and wires (paths) must be sized accordingly. Photo by Tom Monroe.

Wires are the paths that carry current to from source to loads. In a car, they're bundled into *harnesses* to facilitate routing. This factory harness uses special molded connectors to attach wires to components or to other harnesses. Photo by Jim Barrett.

started out as a how-to guide for wiring kit cars, you'll find a number of references to VW Beetle and Ford Pinto components—simply because they're two of the most common donor cars used as the basis for kit-car construction. But this in no way limits the application of the procedures to just these two car lines. Nor should it be taken as an endorsement of the original cars or the kits based on them. So, if you're planning to build a late-model stock car, Chevette-based MG replica, Baja bug, Corvair-powered Ferrari look-alike, rat-motor street rod, or what-have-you, don't worry. Electrical theory and wiring principles apply equally well to your chosen project.

AN OVERVIEW OF THE ELECTRICAL SYSTEM

Someone once said a car needs only two things to run—fuel and electricity. Well, OK, three things, if you count air. But the point is this: A car depends on only these three things for basic operation.

"How can that be?" you ask. "Today's cars are so complicated." Well, compare an old car with a modern one. If you look beyond the obvious, what do you find? The basics are the same. Nowadays, it's only the application that's more com-

plicated. Take electricity.

The essential electrical components in old cars and new ones are practically the same—ignition, starter, generator, battery and lights. But today's cars also come equipped with electric wipers, electric gauges, electric windows—even electric mirrors. Most of today's extras are just that—extras. But all these components have one thing in common. They all need electricity to operate.

So, before getting into the specifics of electrical theory and how to wire your car, let's take a brief overview of the car's electrical system. All the components mentioned here are discussed in detail in their respective chapters.

The electrical system can be divided into three major parts: electrical *sources,* electrical *loads*—users of electricity— and electrical *paths*.

Electrical sources consist of the *battery,* which stores electricity for starting the engine, and the *generator* or *alternator,* which provides electricity when the engine is running. Except in cases of extreme overload, a correctly adjusted charging system will produce enough power to operate all the electrical devices in the car, with enough extra to recharge the battery.

Electrical loads include all the de-

vices on the car that require electricity for operation. Some examples are the ignition system, windshield-wiper motor, heater-blower motor, horn, radio or tape deck, and lights. By the way, don't underestimate the electrical loads created by lights. It's common for a car to have more lights than all the other loads combined—20 to 30 lights can be found in a modern car. And, in some race-car applications, such as rally cars, off-road racers, and IMSA GTO and GTP cars, the electrical load from driving lights can be quadruple that of conventional lighting systems.

Electrical paths include wires, of course. But a car's steel body and frame are also paths. They're used as the *return path,* or *ground,* between loads and the battery. And the various switches in the system can be included in the path category. Switches are located in the electrical path to act as a sort of gate, permitting electricity to pass or to hold it back, as required.

Together, these three parts form electrical *circuits,* which enable electricity to perform useful work. With this brief overview in mind, turn the page and begin your study of electrical theory.

Electrical Theory

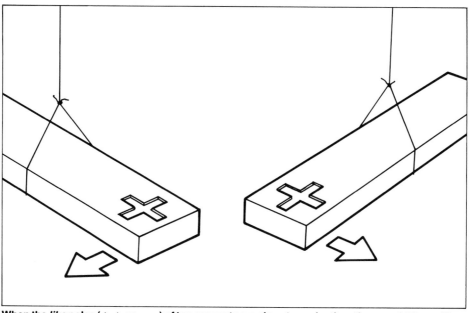

When the *like* poles (+ + or − −) of two magnets are close to each other, they repel. The *unlike* poles (+ and −) attract each other. This phenomenon is the basic key to understanding how electricity works.

It's remotely possible for you to successfully wire a car without any idea whatsoever about electrical theory. *But don't try it.*

Why? It would be similar to taking a trip without a road map. You may have an idea of where you want to go, but no clear-cut idea of how to get there. Or, perhaps more important, you don't know what's involved in making the trip. So, I strongly suggest you take the time to learn basic electrical theory before you try to wire your car.

START WITH THE BASICS

Based on many conversations I've had with people over the years, I've come to the conclusion that, of all the elements that go into a car, electricity is, by a wide margin, the most misunderstood. And not just by the layman, either. Electricity is a hang-up for technicians and engineers, as well.

I've also concluded that a great deal of this misunderstanding stems from the fact that electricity is invisible and formless, so its nature can be perceived only indirectly. Another reason for the misunderstanding is that electricity is not often taught well in many schools, possibly because the instructors don't fully understand it, either.

In the following section, I portray electricity as a phenomenon—which it certainly is—that can be reckoned with and worked with.

So, what is electricity? What is this force that sparks the sparkplugs and lights the lights? Is it some kind of magic that only an engineer or scientist can understand? No!

Webster's New Collegiate Dictionary puts it like this: "Electricity is a fundamental entity of nature consisting of positive and negative kinds...usually utilized in the form of electrical currents."

Mysterious, maybe, but not magic. Let's try to improve on Webster:
● Electricity has positive and negative charges.
● It is an invisible force.
● It can be controlled.
● It can do work.

● It can flow in a current from place to place.
● It can flow only in a completed circuit.
● It can be "stored" in a battery.

Positive & Negative—In grade-school science, you probably learned that magnets are electrical in nature, with a *north pole* and a *south pole*. On some magnets, *north* is marked + (positive) and *south* is marked − (negative). Put two magnets close to each other and you'll see that like poles (+ + or − −) repel each other and unlike poles (+ −) attract each other.

This separation into positives and negatives is called *polarity*. It is the first important point about electricity, which will become clearer later. For now, just remember positive, negative, and the attraction between them.

The Invisible Force—You can't see electricity because it operates at an atomic level. All matter is composed of atoms. They're so tiny it would take millions of them to equal the width of the period at the end of this sentence. Atoms are composed of even smaller par-

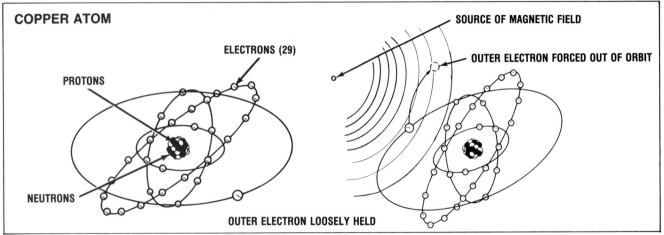

COPPER ATOM

PROTONS

ELECTRONS (29)

NEUTRONS

SOURCE OF MAGNETIC FIELD

OUTER ELECTRON FORCED OUT OF ORBIT

OUTER ELECTRON LOOSELY HELD

Principle of electrical flow depends on fact that atoms of some materials have loosely held electrons in their outer orbits. When loosely held electron is forced from its outer orbit, it is free to "hop" to another atom. Drawing courtesy of Chrysler Corp.

ticles—protons, neutrons and electrons among them. The particles we're concerned with are the *electrons*.

Electrons travel in circular paths around their atom's nucleus, like planets around the sun. In some materials—copper wire for instance—the outermost electrons aren't held too strongly. They're free to hop randomly from one atom to another. Each time an extra electron hops into an atom, it pops another one loose. Because the activity of these free electrons is random, there's nothing directional about their movements.

There are two devices in a car's electrical system that are sources of excess free electrons: *battery* and *generator*. At one terminal there's an excess of free electrons, and at the other terminal there's not enough.

If you attach a copper wire between the two terminals of a battery or generator, free electrons begin to push into the copper wire. The random movement of the copper's free electrons begins to take on a pattern. With *pressure* (voltage) behind and a void ahead, the electrons move in a more-or-less orderly fashion in one direction. This directional flow of electrons in the wire is called electrical *current*. Voltage is discussed later.

Conductors, Resistors & Insulators—
As mentioned earlier, the outermost electrons in copper atoms are held relatively loosely. Silver is another metal

with loosely held electrons. Materials such as these are called *conductors* because they freely conduct or carry electron movement.

Other metals conduct electricity, but not as well. In these metals, when a free electron hits an atom, some of its energy gets absorbed. So when the next electron is freed, it doesn't have as much energy as the first one did. This slowing of electron movement is called *resistance,* and materials in which it occurs are called *resistors*. They conduct electricity, but tend to *resist* its movement. Examples of resistance metals are nickel, iron and nichrome.

Materials whose outermost electrons are firmly held are called *insulators*. Because there aren't any, or not many, free electrons available, insulators can't conduct electron flow—note the difference between a resistor and an insulator. Insulators include air, rubber, glass, wood, bakelite, cotton, and of course, the plastic insulation on electrical wire.

So, metals are capable of conducting electricity, although some act as resistors. Non-metals generally act as insulators.

UNDERSTANDING OHM'S LAW
Like many things in nature, there's a law governing the basic properties of electricity. It's called *Ohm's law*, named for its originator, George S. Ohm, a Ger-

SPECIAL NOTE ON CURRENT FLOW
There are two schools of thought on *which* direction electrons flow in a circuit. Traditional theory contends that current flow is from *positive (+)* to *negative (−)*. That is, excess free electrons flow from a battery's positive terminal, through the circuit, and back to the battery's negative terminal. However, through scientific experiment, this idea has largely been replaced by the modern theory that electrons actually flow from *negative* to *positive*.

Practically speaking, when you wire a car it really doesn't matter which theory you accept. But to avoid confusion, this book assumes that current flow is from *positive* to *negative*. That's the direction favored by tradition, so it is the direction implied by all standard automobile-wiring diagrams. In my opinion, it's also the direction easiest to explain and understand. And the intent here is to explain how to wire a car, not how to pass a graduate class in physics!

man scientist.

Before I get to the formula, here's a simple analogy to help you understand it.

Get a piece of flexible tubing and a bunch of marbles that will comfortably pass through the tubing. Begin by filling the tube with marbles. The marbles

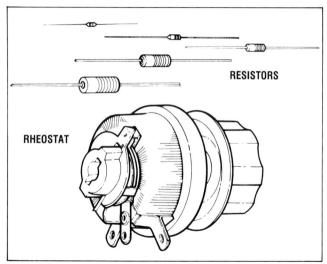

RESISTORS

RHEOSTAT

Resistors come in all sizes and shapes. Small tubular-shaped ones are often used in electronic circuits, such as radios. Larger one shown is a variable resistor, called a *rheostat.* One use of a rheostat is in a headlamp switch to control gauge illumination. Drawing courtesy of Chrysler Corp.

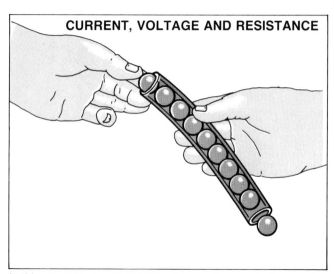

CURRENT, VOLTAGE AND RESISTANCE

In this example, the plastic tube represents a conductor, and the marbles as moving electrons, or *current.* Thumb applying force to push marbles represents *voltage.* Squeezing the tube represents *resistance.*

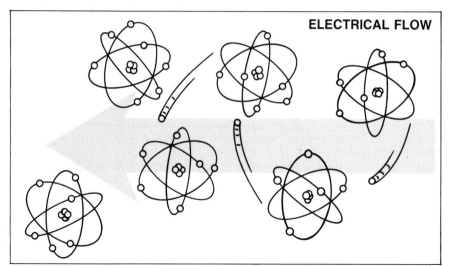

ELECTRICAL FLOW

When a great many electrons begin moving from atom to atom at the same time, the result is electron flow, which we call electricity.

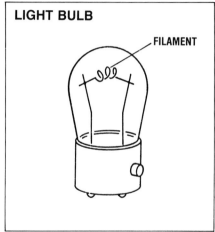

LIGHT BULB

FILAMENT

Filament of common light bulb provides controlled resistance to current flow. Light results when the current flow is great enough to create heat.

represent the movable electrons mentioned in the first section. One marble equals *1 coulomb.* Now push one new marble into the tube. You can readily see that the new marble will push on the one next to it, and so on, until the last marble in the tube pops out. If this took one second, you have a flow rate of *1 ampere,* or *1 coulomb per second.*

What made the marble move? Your thumb. It was the force, or pressure, making the flow happen. The pressure exerted by your thumb represents *voltage.*

Suppose, while you're pushing one marble per second into the tube, a friend grabs the tube and pinches it in a little. Now, you find it's not as easy to push the marbles. The pinched-in section of the tube represents *resistance.* The amount of squeeze is *ohms.* If your friend pinches harder, you have to push on the marbles harder to maintain the same flow rate.

Ohm's law is the electrical law tying your thumb, the marbles and your friend's squeeze together. It works like this: *Thumb = marbles X pinch,* or, *volts = amperes X ohms.* In mathematics, the

customary way to show this equation is:

$$E = I \times R$$

E = volts (pressure)
I = amperes (current flow)
R = ohms (resistance)

Don't worry too much about understanding Ohm's law just yet. It is given here only to introduce you to the three important measurements in electricity: volts, amperes and ohms. Later, after I've discussed the work electricity does and something about circuit theory,

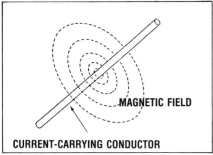

CURRENT-CARRYING CONDUCTOR

When a wire conducts current it is surrounded by a magnetic field, represented here by a series of concentric rings surrounding the wire. This illustrates the basic principle of *electromagnetism*. Drawing courtesy of Chrysler Corp.

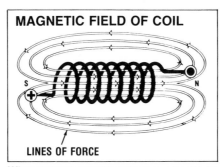

MAGNETIC FIELD OF COIL

LINES OF FORCE

When current-carrying wire is wrapped into a coil, magnetic field is concentrated enough to do work, such as in a motor, solenoid or relay. Like a simple bar magnet, coil creates lines of force, and has polarity (North and South poles). Drawing courtesy of Chrysler Corp.

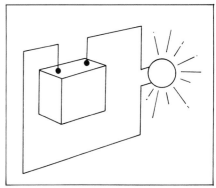

For a circuit to be complete, it has to have *at least* three elements: a source of voltage (battery or generator) a load (lamp, motor or other current-using device) and a path (wires, cables or printed circuits).

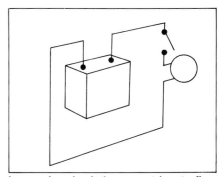

In a *series circuit* the current has to flow through each component in turn as it follows the electrical path. Note here that current cannot flow through lamp because switch is open.

I'll come back to Ohm's law and its applications.

WORK & ELECTRICITY

In a car, electricity performs work in one of two ways: It creates *heat* or it creates *magnetism*. All *electrical* components work on one of these two principles. *Electronic* components like radios and some voltage regulators work on other principles. But you don't have to deal with their inner workings directly when you wire a car, so these principles are not covered in this book. Just one word of advice when dealing with electronic components: Make sure they're hooked up correctly!

Heat is created by resistance. Think of it this way: When electrical current moves through the wires and other conductors, they either permit easy passage or they try to restrict current flow. Restricting flow isn't exactly friction, but you can think of it that way. When the electricity rubs its way through a resistance, the "friction" makes heat. It's similar to sliding down a rope. Your hands get hot from the friction.

One of the main automotive uses of electrical resistance is light bulbs. The bulb filaments get so hot they emit light. It's like when you use a coat hanger to roast hot dogs over a campfire. After a while, the coat-hanger wire holding the hot dog gets hot and begins to give off light.

Other automotive applications of electrical resistance are the coils inside sending units of some instrument-panel gauges, the rear-window defogger, the cigarette lighter and the clicker mechanism in turn-signal flashers.

Other automotive components convert electricity into magnetism. When current flows through a wire, there's a small magnetic field around it. It's almost too small to notice. But if you wind a lot of wire into a coil, the magnetic field is concentrated enough to do work.

Electric motors work by magnetism. Basically, a magnetic field is generated by coils in the motor, which act on an armature that drives the motor shaft. The principles of electric-motor operation are covered in detail on pages 56-58. Electric motors used in cars include the wiper motor, washer-pump motor, starter motor, electric fuel pump, heater-fan motor, and motors for power windows and seats.

Electric magnets also play a role in the operation of relays. Here, magnetic force is used to open and close electrical circuits. A good example is the relay that operates some car horns. There's also a powerful relay in the starting motor. These relays consist of a coil of wire wound around a *core*. For more on relays and how they work, see page 33.

CIRCUITS

Up to this point, you've read about how electricity flows when voltage provides the pressure to move electrons through a conductor. This section describes the actual paths electrons take. These are called *circuits,* and they have certain specific characteristics you need to know about when you're wiring a car.

Simple Circuit—Start with a battery. Hook a wire to the positive terminal. Hook another wire to the negative terminal. Now attach the free ends of the wires to the terminals of a lamp. The lamp lights. Let's see what you've done. You've provided a path from the battery positive terminal to the lamp, through the lamp filament, and back to the battery negative terminal. This is a *complete* circuit.

The simplest kind of circuit includes a *battery* (electrical source), a *load* and some *wire* (electrical path). In the example, the lamp is the load because it's what is using up the battery's current to do work. In this case, the work is to provide light.

The single most important characteristic of a simple circuit, or any circuit, is

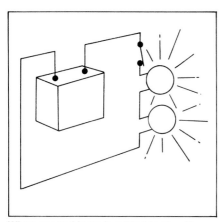

When second lamp is added to circuit, current has to pass through switch and both lamps. Switch and lamps are *in series* with each other. If one bulb blows, both lamps go out.

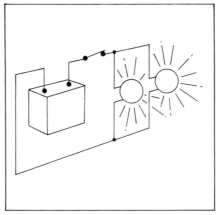

In a *parallel circuit* current has a choice; it can flow through each lamp separately. If one bulb blows, other lamp stays on.

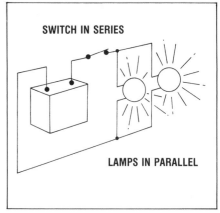

In a *series/parallel* circuit, current can flow through each lamp separately, but has to flow through switch in series to get to either lamp. One switch controls both lamps simultaneously. This is probably the most common circuit used in automotive electrical systems.

that it must start at the battery positive terminal and end at the battery negative terminal. Without this set of connections, current won't flow.

If you want to turn off the lamp, you must break the circuit. You could disconnect any of the four connections to do this. Disconnecting a wire to turn off a lamp isn't much trouble in a table-top experiment, but it would be intolerable in a car. So you can see that the simplest circuit is also an impractical circuit. Clearly, you need some sort of control to turn the lamp on and off. This brings up a second kind of circuit—the *series circuit*.

Series Circuit—Instead of disconnecting one of the wires to turn off the lamp, simply cut the wire between the battery positive terminal and the lamp. This certainly breaks the flow of current. Then install a switch between the two cut ends. With the switch in place, you can *make* and *break* the circuit any time you want by operating the switch.

Now, electricity must flow first through the switch, then the lamp, on its way from the battery positive terminal to the battery negative terminal. Electricity must pass through a *series* of components, one at a time. That's why this is called a *series circuit*.

Suppose you want your circuit to have two lamps operated by one switch. How could you wire the second one in?

You could cut the wire again and attach the new cut ends to the second lamp. Now, current flows through the switch, through the original lamp, through the second lamp and back to the battery. Notice again that current must pass through one component to get to the next one.

The advantage of this kind of circuit is obvious. You don't need any extra wire to add the second lamp. But, there's a big disadvantage. If one of the bulbs burns out, the filament inside actually burns in half, breaking the circuit. Current can't flow when the circuit is broken, so the other lamp goes out, too. You may remember the old-fashioned Christmas-tree lights that worked this way. You actually had to test each bulb separately to find the bad one.

There's another disadvantage to the series circuit, too. It happens when you put a high-resistance component in series with a low-resistance one. Current flow through the combined components may not be sufficient to permit the low-resistance component to operate. And, the order in which they're wired doesn't matter. High-resistance components simply choke current flow. Later, when I get back to the applications of Ohm's law, the reasons for this should become clear.

Parallel Circuit—There's another way to wire a second lamp into your circuit.

This time, you won't disturb the original wiring at all. Instead, add two more wires and the second lamp. Now, current flowing from the battery goes through the switch to the first lamp—and to the second lamp—at the same time. Current can flow through both filaments at the same time and back to the battery. There are two *parallel* paths for current to follow.

In a *parallel circuit,* one bulb keeps working even if the other bulb burns out. The parallel circuit requires extra wire, though.

Series or Parallel?—How do you decide when to use a series circuit and when to use a parallel one? You must ask yourself one question: "Do I want each component to get current independently, or do I want the current to flow through one before it gets to the other?" The analysis may surprise you.

Think about this for a minute: What good would a switch do connected in parallel to the lamp in your circuit? None. Battery current would be available at the switch and lamp all the time. Any time the battery is connected to the circuit, the lamp will light, regardless of switch position. The switch wouldn't control the lamp at all. For this reason, be sure the switch is always wired in series with the load it is supposed to control.

As mentioned earlier, one of the disadvantages of putting loads in series is

Ground stud or ground bolt is used to provide common grounding point for two or more components. Here, taillamps mounted on fiberglass car body are all grounded to car's metal frame at this one point.

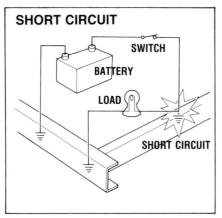

SHORT CIRCUIT

SWITCH
BATTERY
LOAD
SHORT CIRCUIT

Short circuit occurs when current goes directly to ground, bypassing load. Shorts are caused when bare copper conductor accidentally contacts adjacent conductor or grounded metal part of car. Resulting high current flow burns up wire, possibly causing serious electrical fire.

that if one load burns out—opens the circuit—the others quit working, too. So, make sure the loads are in parallel to each other. In many cases, you'll have a combination *series-parallel circuit*. That is, the loads are parallel to each other, but the loads together are in series with the switch controlling them. A car's headlamp circuit is a good example. Both the right and left lamp are operated by the same switch. But if one lamp burns out, the other one will stay on.

It's also possible to control one load from several places. For example, you use a door-actuated switch at each car door to control a single dome lamp. Here, the switches are wired in parallel to each other, but in series with one load. Each switch permits current to flow through the lamp, regardless of the position of the other switch or switches.

Single-Wire Circuits—Up to now, I've described several simple circuits involving a battery, switch, load and some wire. In every case, a second wire carried the current from the load back to the battery. Now for the shock—pun intended. Circuits in cars aren't wired this way. Most use only one wire to carry current from the battery to the load. "But," you might say, "the circuit needs that second wire to get the current back to the battery." Wrong. Running a second wire from each load back to the battery would take a lot of wire, so the car's

sheet-metal body and chassis are used as a common return path (ground) from the load back to the battery.

One battery terminal—normally the negative—is attached directly to some metal part of the car—the engine, chassis or metal body. When engine, chassis and metal body are not connected directly, you'll find large *ground straps* joining these components to form one large, continuous ground. Each load is connected to the engine, chassis or body. Usually the load component's entire case or body forms the ground terminal.

Because modern cars have the negative battery cable attached to the chassis, these systems are called *negative-ground* systems. Some earlier American cars and many imports have a *positive-ground* system. This is especially important to remember because you can't install components with a negative-ground orientation into a vehicle with a positive-ground system.

Before going on, make sure you understand how the single-wire circuit works. Remember—one wire supplies current to the load, while the car's metal structure acts as the second wire to complete the current path from the load back to the battery. It's easy to overlook the metal body and chassis as conductors because they don't look like wires.

Short Circuits—Unlike the other circuits already mentioned, the *short circuit*

is not one you'll find intentionally engineered into a car. Instead, the short circuit is an accidental flaw that works its way into an electrical system.

The following section on the applications of Ohm's law, explains the electrical theory behind a short. For now, let's look at the mechanical aspects.

Consider this example: The wire between the battery positive terminal and a load is accidentally grounded. In a car, this could be caused by a pinched or chafed wire. When a wire's insulation is cut, the exposed copper conductor may touch a grounded metal part of the car. Current goes directly to ground, bypassing the load entirely.

So, what happens? The wire may get hot—maybe even start to smoke. Without a load in the circuit, the current meets little or no resistance, so it races through the wire fast enough to create heat. Soon, the insulation starts to burn and the wire melts. This is one reason why *all* electrical circuits in a car are protected with *fuses* or *circuit breakers*.

Circuit Protection—A fuse is designed so it is the weakest link in a circuit. Normal current flow doesn't bother the fuse element at all. But, if flow (amperage) becomes too great, the fuse element gets hot and melts. The circuit becomes *open* and current flow stops. Fuses melt, or *blow*, at a current flow greater than the normal expected flow, but lower than the flow necessary to melt the wires.

This is why it's important to use correctly rated fuses. If you install one that's rated too low, a normal load may blow it. But, if the fuse is rated too high, it may not blow as it should during a short. This could cause the wires to become the weakest link, burning them up.

Keep in mind that a blown fuse is a *symptom* of a circuit problem. Fuses rarely cause problems themselves.

Fuses are rated by the maximum amperage they can carry before blowing. Some fuses have the number stamped on the fuse body. Others are color-coded according to amperes. Typical fuse ratings for automotive use include 5, 8 and 16 amp. Remembering what was said earlier about amperes being a flow rate, you'll see that 16 amp is twice the flow of 8 amp.

Fuses aren't the only circuit-pro-

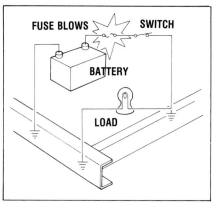

In good circuit design, each circuit is protected by a fuse. Fuse is sized to melt at a lower current flow than wire it protects. In other words, during excessive current draw, fuse acts as weakest "link" in electrical "chain."

12-volt battery powers lamp that draws 2 amp. Applying Ohm's law $(E = I \times R)$, you find that resistance in circuit is 6 ohms.

tection devices used. There are also *mechanical circuit breakers* and *fusible links*. These do about the same thing as a fuse—they open the circuit if current flow is dangerously close to damaging the wiring. These devices are discussed in detail on pages 34-37.

Hot Circuits—So far, several kinds of circuits have been described—series circuits, parallel circuits, and the single-wire circuit in which the car's metal structure acts as a ground.

But there's another way to categorize circuits. A circuit that is connected to the battery positive (B+) terminal is said to have a *voltage potential*. This means that if you want to connect a load to the circuit, electricity is available to make the load function. The common, or non-technical, name for a circuit such as this is *hot*. Some people also call them *live* circuits, which is why a circuit with no voltage potential is often called *dead*.

There are three variations of hot circuits normally used in a car. The first starts directly at the battery positive terminal. Because this circuit is always hot, it's called a *battery circuit* or *B + circuit*. You must be extra careful around a B + circuit because it always has voltage potential.

Typical B + circuits are the emergency flasher, horn (on most cars), dome lamps and cigarette lighter. All these loads must be operable, even when the ignition switch is turned off. The B +

circuit also provides current to the ignition switch for further distribution.

The second kind of hot circuit starts at the ignition switch. Because this kind of circuit is hot only when the ignition switch is on, it is called an *IGN + circuit* or just an *IGN circuit*. You may also see this referred to as an *ignition feed circuit*. Most electrically operated loads get their current by way of an IGN feed. Examples are the ignition coil and the instrument gauges.

The third kind of hot circuit starts at a switch and goes to the load. These circuits are hot only when the switch is turned on. We usually refer to them by the name of the load: *headlamp circuit, turn-signal circuit, back-up-lamp circuit* and so on.

APPLICATION OF OHM'S LAW

Ohm's law has many practical applications in all phases of automotive wiring—circuit design, component selection and troubleshooting—because it allows you to determine the third variable in a circuit if the other two are known. Ohm's law in formula form is: E *(voltage)* $= I$ *(amperes)* $\times R$ *(resistance)*. Depending on which two variables are known, and which variable you're looking for, the formula can also be written: $R = E \div I$, or $I = E \div R$.

The following examples show how Ohm's law is used to determine these variables in both a *series circuit* and a

parallel circuit. In actual practice, you can use one or more of the following test instruments to take measurements and make the calculations. If you know how to use a *voltmeter, ammeter* and *ohmmeter*, you can set up simple experiments to verify the following examples.

Series Circuit—From the description of circuits, you should already know you need a battery, a load and some wire to make a circuit. This example uses a conventional 12-volt car battery and a typical parking-lamp bulb that draws 2 amp, wired in series.

By substituting the known values into the formula, $R = E \div I$, you get: E *(12 volts)* $\div I$ *(2 amp)* $= R$ *(6 ohms)*. So, you find that the bulb has 6 ohms of resistance. This could be verified by using an ohmmeter, *after* disconnecting the battery.

Now, suppose you know the voltage is 12 volts and the resistance of a light bulb is 0.5 ohm. What amperage will flow in this circuit? Using the formula, $I = E \div R$, you get: E *(12 volts)* $\div R$ *(0.5 ohms)* $= I$ *(24 amp)*.

You can deduct an interesting fact from these two examples. The high current user in the second example—the one that flows 24 amp—has a much lower resistance than the low current user in the first example. This is because resistance (ohms) is inversely proportional to amperage, as stated in Ohm's law.

Now, let's see what happens in a short circuit, using Ohm's law. Here, the resistance of the load—the accidental short—is just about negligible. Because no wire has absolutely zero resistance, suppose the resistance in the shorted wire is 0.001 ohm. Voltage remains at 12 volts. What amperage will flow in this short circuit? E *(12 volts)* $\div$ *(0.001 ohm)* $= I$ *(12,000 amp)*. With 12,000 amp racing through the wire in an instant, it's no wonder the wire will melt.

Before leaving series circuits, let's see what happens when there's more than one resistance in a circuit.

Keeping the 12-volt battery and 6-ohm parking-lamp bulb from the first example, let's add a second lamp with 18 ohms resistance.

In a series circuit, the total circuit resistance is the total of all the separate resistances added together. In this example, total resistance is: *6 + 18 = 24 ohms*. Applying Ohm's law, we find: *E (12 volts) ÷ R (24 ohms) = I (0.5 amp)*.

From the earlier discussion of series circuits, you know that when a high-resistance component and a low-resistance component are wired in series, the current flow may not be enough to operate the low-resistance component. This example shows why. At 12 volts, a current flow of only 0.5 amp is not enough to light a 6-ohm lamp that needs 2 amp to operate. At best, current flows through the low-resistance component as though it weren't there. There may be sufficient flow to operate the high-resistance component. In this case, one lamp would light and the other would simply act as a wire. Front side-marker lamps (high resistance) that blink opposite to the front turn signals (lower resistance) use this principle.

Voltage Drop—Let's take a brief look at a phenomenon known as a *voltage drop*. This is the term used for the portion of available voltage that's used up by any particular resistance in a circuit. I will use a 2-ohm and 4-ohm lamp for the example and will discuss each lamp separately.

First, calculate system amperage: *E (12 volts) ÷ R (2 + 4 = 6 ohms) = I (2 amp)*. This means that everywhere in the circuit, you'll find a flow of 2 amp. Now, determine voltage drop across the 2-ohm lamp first: Applying Ohm's law, you get: *I (2 amp) X R (2 ohms) = E (4 volts)*.

You can see that 4 volts are "used up" at the 2-ohm lamp. This is called the *voltage drop* across the resistance. If you put the leads of a voltmeter on either side of the lamp, it will read 4 volts.

Do the same calculation for the 4-ohm lamp and you get a voltage drop of 8 volts. Add the voltage drops of both lamps: *4 + 8 = 12 volts*. Under Ohm's law, if you add all voltage drops in a series circuit, they will total the available voltage, which is battery voltage, or source voltage.

Voltage drop can happen in a circuit with just one apparent load, such as a headlamp. In a circuit with just one headlamp, the headlamp is the only obvious place for a voltage drop. But, what if a terminal connection loosens or gets corroded? What if a wire gets cut partially

through, thus adding a voltage drop?

Any one of these can effectively reduce the size of the path the current must flow through. It sounds like a resistance, doesn't it? That's exactly what it is. And it's in series with the headlamp. You don't need Ohm's law for this. You already know what happens when there are two resistances in series. Amperage goes down. Each component—in this case, the headlamp—gets less voltage to work with. The headlamp dims, due to a voltage drop somewhere else in the circuit.

The applications of Ohm's law for a series circuit can be summed up in three statements. First, current (amperes) is the same through each resistance. Voltage drop across a resistance depends on the amount of resistance. Finally, the sum of voltage drops in a circuit equal the source voltage.

Parallel Circuit—For a parallel circuit, you need a battery, some wire, and at least two loads, each receiving full voltage independent of the other.

In this example, the two lamps are rated at 6 ohms and 3 ohms. Battery voltage is still 12 volts. A quick glance at the circuit reveals why there are 12 volts available at each lamp—current doesn't have to pass through anything but wire to get to each lamp.

Applying Ohm's law for the 6-ohm lamp: *E (12 volts) ÷ R (6 ohms) = I (2 amp)*.

Applying Ohm's law to the 3-ohm lamp: *E (12 volts) ÷ R (3 ohms) = I (4 amp)*.

Total current in the system is: *2 amp + 4 amp = 6 amp*. Now that you know system current is 6 amp and battery voltage is 12 volts, you can apply Ohm's law again to find total system resistance: *E (12 volts) ÷ I (6 amp) = 2 ohms*.

The conclusion here is that a 6-ohm lamp and a 3-ohm lamp in parallel offer a combined resistance as though there were a single 2-ohm lamp.

There's another formula for figuring the equivalent resistance of any number of resistances in parallel. It's a little mind-boggling to look at, and I apologize for it.

$$R = \cfrac{1}{\cfrac{1}{r_1} + \cfrac{1}{\cfrac{1}{r_2}} + \cfrac{1}{r_3}}$$

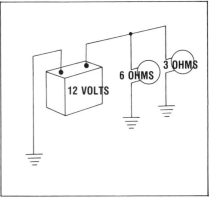

To determine current flow in parallel circuit involving 6-ohm lamp and 3-ohm lamp, apply Ohm's law for each branch separately. Now, add amperage of both lamps and apply Ohm's law again to determine total circuit resistance. Surprisingly, total resistance (2 ohms) is lower than either one of individual lamps.

Where:
R = equivalent resistance
r_1 = first resistance
r_2 = second resistance
r_3 = third resistance, and so on.

I'll solve just one example to show you how it's done. And I'll keep it simple:

r_1 = 1 ohm
r_2 = 2 ohms
r_3 = 3 ohms

$$R = \cfrac{1}{\cfrac{1}{1} + \cfrac{1}{\cfrac{1}{2}} + \cfrac{1}{3}} = \cfrac{1}{\cfrac{6}{6} + \cfrac{3}{6} + \cfrac{2}{6}}$$

$$R = \frac{1}{1\frac{5}{6}} = \frac{1}{1.83} = 0.55 \text{ ohm}$$

As you can see, the combined resistance (about 1/2 ohm) is lower than any of the individual resistances. This is always true of parallel resistances.

SUMMARY

This chapter covered the basics of automobile electrical theory. In later chapters, some of these principles are applied to how components work, circuit design and test procedures you can perform on your wiring installation. Terms and principles explained in this chapter are not redefined later, so be sure you understand what's here before moving on.

Charging System

This is a 2.8 litre, 60° Chevy V6 setup for a Class 7 off-road racing S-10 pickup truck. You can't miss noticing large 90-amp alternator, used to power flamethrowers required for night racing. Photo by Tom Monroe.

The power source for all electrically operated devices on a car is referred to as the *charging system*. This system consists of the *battery, generator (alternator)* and *voltage regulator*. Unless specified, I'll use the term *generator* for generator and alternator.

During normal operation of a car, most, if not all, electrical needs are supplied by the *generator* or *alternator*. The one major exception is during engine start-up, when the *battery* supplies all the current to turn the starter motor and operate the ignition system. Also, there are times while driving when the battery supplements generator output to handle excessive loads. An extreme example is when the engine of a standard car is idling in traffic, with ignition, heater blower, lights, wiper motor, rear-window defogger and stereo all demanding electricity at the same time.

It's obvious, then, that the battery plays an important role in the car's electrical system.

BATTERY

It's helpful to think of the battery as a place to store electricity. It actually stores chemicals, not electricity, but the former idea is too well established to shake off. For this reason, batteries are often called *storage batteries* or *storage cells*.

However, batteries depend on a *chemical reaction* to produce electricity, as described below. This principle was discovered by Alessandro Volta in the late 1700s. The term *volt* is derived from his name.

In Volta's battery, two *dissimilar metals* are suspended in a salt or acid solution, called *electrolyte*. The electrolyte solution reacts with one of the metals more than the other one. The more-active metal loses electrons and becomes *positively charged*. The less-active metal collects electrons and becomes *negatively charged*. The amount of positive or negative charge is called *potential*. The difference in potential between the unlike

charges creates a pressure that causes current to flow whenever a path is provided, such as an electrical load wired to the positive and negative terminals of the battery.

CHARGE & DISCHARGE CYCLES

There's one important feature of a car battery that distinguishes it from a common dry-cell battery, such as those used in flashlights: A car battery can be discharged and recharged many times. Standard dry-cell batteries can't be recharged because the chemical materials are depleted in the process of making electricity.

The discharge process works like this: Let's begin with a fully charged automotive battery. True to Volta's principle, it contains two unlike metals, in this case *lead* and *lead peroxide*. These are suspended in an electrolyte solution of water and sulfuric acid. When a load is connected to the battery terminals, a chemical reaction inside the battery causes cur-

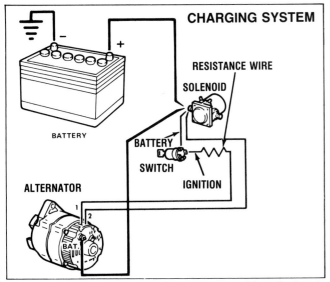

CHARGING SYSTEM

During normal operation, charging system provides all of car's electrical needs. Generator or alternator, controlled by voltage regulator, should generate enough voltage to operate car's loads as well as to recharge battery. Drawing courtesy of American Motors Corp.

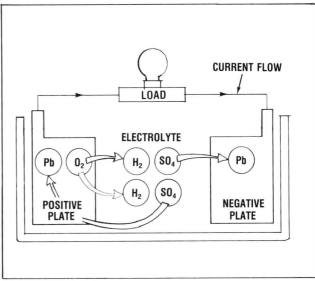

Car battery uses electrodes of lead and lead peroxide. During discharge process, when current flows from battery, both electrodes slowly convert to lead sulphate. Charging reverses this process. Drawing courtesy of Chrysler Corp.

rent to flow, discharging the battery. Eventually, both metals will become *lead sulphate,* and the electrolyte plain water. At this point, no more electricity can be made, so current flow stops. The battery is now fully discharged.

The *charging process* reverses these conditions. Running an electrical current through the battery causes the metals to become dissimilar again, and changes the electrolyte back to sulfuric acid. When the charging process is complete, the battery is again capable of providing electricity. In practice, however, a car battery never becomes fully discharged under normal operating conditions. It loses part of its charge when called upon to start the car or operate electrical devices. This loss is replaced by the generator when the engine is running.

With proper care, a car battery can undergo continuous discharge-charge cycles over many years before wearing out. See page 18 for a discussion on how a generator works to recharge a battery.

WARNING: During the charging process, all batteries generate some hydrogen gas—even the "sealed" types. The gas is highly flammable. If ignited by a stray spark or open flame, it could explode, causing personal injury or damage to the car.

BATTERY CONSTRUCTION

An automotive battery consists of several individual compartments, called *cells.* There are three cells in a 6-volt car battery, six cells in a 12-volt car battery. Each cell provides slightly more than 2 volts, regardless of size.

Each cell contains the two dissimilar metals mentioned earlier—lead and lead peroxide. These metals are formed into flat, porous *plates* that stand vertically in the cell. The number of plates, and the surface area of each, determine a battery's capacity to generate electrical current, which is rated in *amp hours,* page 17. More plates, larger plates, or both, will provide more capacity. But, remember that each cell delivers only 2 volts, regardless of size.

Each cell starts with a *negative plate,* followed by a *positive plate,* followed by another negative plate, and so on, finally ending with a negative plate. This means each cell has one more negative plate than it does positive plates.

At the top of the battery, all negative plates are welded to a common *plate strap;* all positive plates are welded to another plate strap. Perforated *separators* prevent the plates from touching each other, but allow electrolyte to pass easily between them. Individual cells are

separated by a solid partition. At the top of the battery, the plate straps are joined in series from one cell to the next. At the bottom, the plates sit on ribs that provide space below the plates for sediment to collect without shorting the plates. At each end of the battery, the final plate strap is formed into a post that protrudes

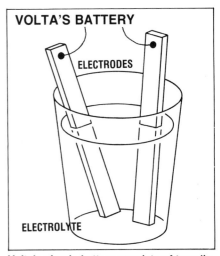

VOLTA'S BATTERY

Volta's simple battery consists of two dissimilar metals, called *electrodes,* suspended in fluid called *electrolyte.* A voltage potential appears across wires attached to electrodes. Car batteries work on this principle.

BATTERY COMPONENTS

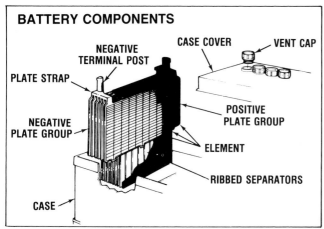

Electrodes in car battery are formed into plates suspended in solution of water and sulfuric acid, called *electrolyte.* Negative and positive plates are alternated in each cell. Drawing courtesy of Chrysler Corp.

BATTERY GRID CONSTRUCTION

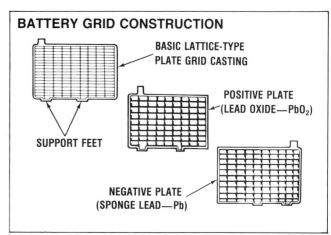

Battery plates are formed into a grid pattern to provide maximum exposure to the electrolyte. Drawing courtesy of Chrysler Corp.

LOW-MAINTENANCE BATTERY

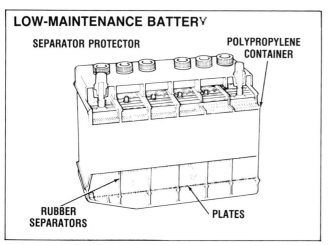

Features of long-life, low-maintenance battery include heavy-duty plates, polypropylene plastic container, rubber separators and separator protectors molded into plastic top. One shown here has caps to check electrolyte level and add water as needed. Other types are "sealed for life."

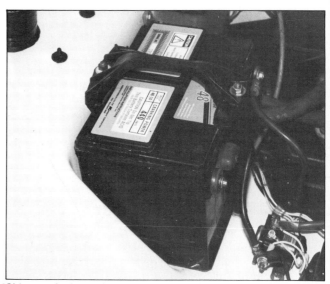

Side terminals on low-maintenance battery ensure that fumes emitted by battery do not corrode terminals. Because all car batteries create some hydrogen gas during normal operation, provide good ventilation.

through the top or side of the case, forming positive and negative *terminals.*

On post-type batteries, the terminal posts are tapered, with the positive post being larger than the negative—11/16 in. and 5/8 in., respectively. The size difference makes it less likely for you to attach the cables to the wrong posts. Tapered posts require a clamp-on cable terminal. Good-quality cable clamps are sized accordingly; cheap clamps are not.

Many battery manufacturers are offering batteries with the terminals at the side of the case, rather than on top. Side terminals are internally threaded to accept a mating cable with a built-in bolt.

Regardless of terminal construction, *terminal polarity* is always identified, either on the terminal or the battery case. Positive terminals are identified with P, POS, +, or red paint. Negative terminals are N, NEG, −, or paint of some color

other than red—often black.

Some batteries have *removable caps* to permit periodic checking of electrolyte level. Distilled water is usually recommended for topping off low cells. *Never add electrolyte.*

Many new batteries are "sealed for life." The chemical composition inside the battery is such that very little hydrogen gas is given off at normal charging voltages, so little of the electrolyte is consumed over the life of the battery. Sealed batteries are preferred because they eliminate the need for periodic maintenance. The types with side terminals are designed so the terminals and battery cables are located below the battery's vent hole, reducing terminal corrosion.

Gel cell batteries are specially constructed to withstand severe shock, and are good for such applications as off-road racing. The plates on conventional batteries can short out when subjected to excessive vibration or shock.

BATTERY RATINGS

Two different systems have been used to rate battery strength, or output ability. The present system was developed during the 1970s.

In the earlier system, developed by the Society of Automotive Engineers (SAE), a battery was rated in *amp-hours.*

Amp-hours are determined as follows: With the battery ambient temperature at 80F (27C), determine the maximum continuous amperes the battery can deliver for 20 hours with each individual cell maintaining at least 1.75 volts. Example: *5 continuous amperes X 20 hours = 100 amp-hours.*

The amp-hour rating is a laboratory-oriented standard that's useful only in comparing one battery to another. The rating doesn't translate directly to any useful, real-world application.

In the real world, the scenario may go something like this: You're driving at night, many miles from home, and suddenly the charge indicator tells you the generator just died. Will the unassisted battery be able to power the ignition system and lights long enough to get you home? The present rating system, called *reserve-capacity battery rating,* could tell you.

Reserve-Capacity Rating—By definition, the reserve-capacity rating is the number of minutes a fully charged 12-volt battery at 80F can be discharged at a constant 25 amp to maintain at least 1.75 volts per cell—10.5 volts for the battery. A 75-minute battery would last 75 minutes under these conditions, a 110-minute battery 110 minutes, and so on. Obviously, the higher the rating number, the longer the battery will deliver

adequate current if the generator fails.

To answer the question about that hypothetical dark night, you first add up the amperage requirements of headlamps, taillamps and ignition system. Divide this number into 25 and multiply that answer by the reserve-capacity rating of the battery. This gives the actual number of minutes the battery will last under such conditions.
Example:

Ignition + two headlamps +
two taillamps = total amperes.
Or:

2 amp + (2 X 4.5 amp) + (2 X 0.6 amp)
= 12.2 amp.

25 standard amp ÷ 12. car's amp
= 2 amp (approx.)

2 X 75 minutes (reserve-capacity rating)
= 150 min. = 2-1/2 hrs.

Cold-Crank Rating—Another way of rating a battery is to determine its ability to operate the starter motor under extremely cold conditions. This is called the *cold-crank rating.* When temperature drops, the chemical reaction taking place inside the battery slows down. Using the SAE standard temperature of 80F, a battery will put out 100% of its rated power. At 32F (0C) the available power drops to about 65%; at 0F (-18C) it's down to only 40%.

A battery's cold-crank rating specifies the minimum amperes a fully charged

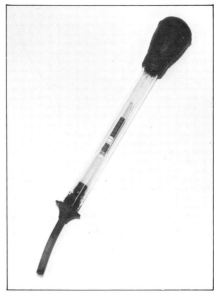

Hydrometer contains weighted float, which sits shallower or deeper in electrolyte sample, depending on specific gravity. Reading on float is taken at point where it protrudes from liquid.

12-volt battery can deliver for 30 seconds at 0F without falling below 7.2 battery volts. Typical cold-crank ratings are 305 amp, 385 amp and 410 amp. It's not uncommon for a starter motor—depending on engine size—to draw at least 300

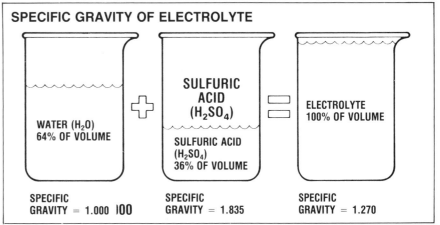

SPECIFIC GRAVITY OF ELECTROLYTE

WATER (H_2O) 64% OF VOLUME

SULFURIC ACID (H_2SO_4)

SULFURIC ACID (H_2SO_4) 36% OF VOLUME

ELECTROLYTE 100% OF VOLUME

SPECIFIC GRAVITY = 1.000

SPECIFIC GRAVITY = 1.835

SPECIFIC GRAVITY = 1.270

Pure water has a specific gravity of 1.000; sulfuric acid a specific gravity of 1.835. By measuring the specific gravity of the electrolyte in a battery, you can determine the ratio between sufuric acid and water, thus the amount of charge. 1.270 is ideal. Drawing courtesy of Chrysler Corp.

amp during cold weather. So, you can see how important cold-crank ratings can be, particularly in Northern climates.

Specific Gravity—Another battery characteristic related to temperature is the *specific gravity* of the electrolyte, made up of water and sulfuric acid. As a battery discharges, the electrolyte solution contains less acid and more water, thus its specific gravity (weight) changes.

Because acid is heavier than water, you can determine the approximate state of the battery's charge by comparing the weight of its electrolyte to the weight of pure water, or its specific gravity. A *battery hydrometer* is used to make this comparison.

Draw an electrolyte sample into the hydrometer and read the specific gravity on the floating scale. Most hydrometers are calibrated for the standard temperature of 80F. Add 0.004 to the hydrometer reading for every 10 degrees above 80F; subtract 0.004 for each 10 degrees below 80F.

Most good hydrometers have both a floating scale and thermometer in the same housing, and a conversion table printed on the barrel. A fully charged battery will read 1.260; a fully discharged one will be in the 1.110—1.120 range.

Battery testing other than reading the specific gravity requires expensive test equipment, but a local battery-service facility can perform these tests if needed.

Instead of being too concerned with further testing of failed batteries, consider eliminating the things that cause batteries to fail. You can contribute to the battery's long, successful life by watching out for these problems:
- Defects in the generating system.
- Overloads (using a too-small battery).
- Dirt and corrosion on battery or cable terminals.
- Damage to lead plates from operating in a constantly undercharged condition.
- Physical damage to the case.
- Under-hood heat (placement near exhaust system).

BATTERY SELECTION

Selecting a suitable battery for your project car is much more than just getting the right price. Once you understand the theory behind battery ratings, you need to know how to translate that theory into a buying decision. The following guidelines will help:

Cranking capacity is the single most important factor. The rule of thumb is to buy a battery with at least 1 amp of cold-crank rating for each cubic inch of engine displacement. This translates to a minimum of 6 amp per 0.1 liter, considering there are approximately 61 cubic inches per liter. For example, a 1.6-liter stock VW engine requires a battery rated at 96 cold-crank amp. Because most replacement batteries are sold for engines with larger displacements, you may have trouble finding a battery rated this low. In other words, you probably won't have to worry about any battery, whatever the price, being inadequate for a stock VW engine. For larger engines, use the above rule.

There's also a rule of thumb for determining *reserve capacity*. Unless you frequently travel in remote areas, choose a battery whose reserve capacity permits you to drive the car with a failed a generator for at least 1 hour.

Then there's the matter of warranty. Because the price of a battery is roughly proportional to its warranty coverage, it's evident that you get basically what you pay for, at least with major brands. Be aware that warranties are prorated against time in service. That is, a 5-year battery that's been used for 2 years may be warranted for 2/5 of the list price, not the full list price. As described below, this can vary.

Also concerning list prices, you may find new batteries sold at discounts considerably below "full lists." This sometimes results in the unusual consequence of your worn-out battery having a *prorated value*—based on full list—that's close to the *discounted price* for a new battery. In such cases—and they're common—the warranty isn't worth much.

GENERATOR

If the battery is the heart of your car's electrical system, the generator is second in importance. Actually, both interact so closely that it's almost impossible to separate their functions entirely. This sec-

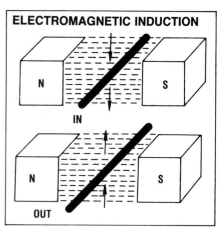

ELECTROMAGNETIC INDUCTION

Moving wire through magnetic field induces voltage into wire. If wire is moved one way, then the other, alternating current flows in wire. Drawing courtesy of Chrysler Corp.

tion explains how the generator provides current for the car's electrical system and how it acts to recharge the battery.

As mentioned in the battery section, all electrical needs are usually supplied by the generator while the engine is running. An exception would be when the current demand of the car's electrical system exceeds generator output. When this happens, the battery supplements generator output. Also, the battery provides all current needed to start the car—that is, cranking the engine and supplying power to the ignition system.

To understand current generation, you need to know about an electrical phenomenon briefly discussed in the theory section—*magnetism*.

A magnet has a *north* and *south* pole. These create an invisible *magnetic field* around the magnet. If you place the north pole of one magnet near the south pole of another, the magnetic field extends from one magnet to the other through the intervening space, or air gap.

If you hold a piece of wire between the two poles at a right angle to the magnetic field, and move the wire up and down, voltage will be induced into the wire. This phenomenon is called *magnetic induction*. The direction the current flows depends on the direction of the magnetic field and the direction the wire is moving. Current flows one way when you move the wire up and the other way when you move it down. Incidentally, you could hold the wire still and move the

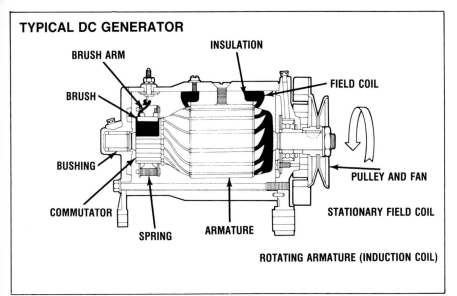

TYPICAL DC GENERATOR

- BRUSH ARM
- INSULATION
- BRUSH
- FIELD COIL
- BUSHING
- PULLEY AND FAN
- COMMUTATOR
- STATIONARY FIELD COIL
- SPRING
- ARMATURE

ROTATING ARMATURE (INDUCTION COIL)

Direct current (DC) generator is no longer installed in new cars. Output was limited at low engine speeds and not satisfactory for high current demands of modern cars. Drawing courtesy of Chrysler Corp.

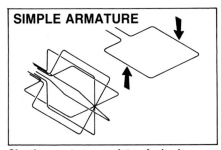

SIMPLE ARMATURE

Simple armature consists of wire loop rotated through magnetic field to induce direct current into wire. Armature of DC generator has large number of loops wrapped around laminated iron core. Drawing courtesy of Chevrolet Motor Divison.

magnets instead—the result would be the same.

To make a useful generator, this induced voltage must be harnessed. That is, *mechanical energy*—moving the wire—must be converted into to *electrical energy*—the induced voltage.

In the moving-wire example, repeatedly moving a wire up and down in a magnetic field causes current to flow one way, then the other. Because the directions alternate, this is known as *alternating current* (AC). In a car, the current always flows in one direction; from the battery to the load. This is known as *direct current* (DC). But, depending on the year of manufacture, your car will have either a *DC generator* or *AC generator (alternator)*. The next two sections describe each in detail.

DC GENERATOR

You should know at the outset that the DC generator is not used in modern cars. Although it is less complex than the AC generator, its output is inadequate for the high current demand of most modern electrical systems, particularly at low engine speeds.

However, the DC generator was used on all cars until Chrysler introduced the AC generator in 1960. So, if you're restoring an older car, you must deal with a DC generator. So will kit-car builders using a VW older than 1973.

In an automobile generator, it would not be practical to rely on permanent magnets to provide the necessary magnetic field. They aren't very strong and they tend to weaken with age. So, the DC generator has a set of electromagnets called *field coils*. These consist of numerous wire windings on iron-pole *shoes,* energized by the car's electrical system. The battery provides the initial current to energize the field, then the generator takes over and "feeds" itself when operating speed is high enough.

In the moving-wire example, a single wire was moved through a magnetic field to produce current. But in a generator, a single wire wouldn't generate enough current to be useful. So, a generator has a large number of individual wire loops wrapped around a laminated iron core called an *armature,* which is turned by a belt driven by the engine. The two ends of each wire loop are attached to copper *commutator bars.* The bars are arranged in a circle at one end of the armature shaft such that the bars for each loop are positioned directly opposite each other, or 180° apart. The commutator bars are insulated from each other and from the armature shaft with pieces of mica.

Two carbon-based *brushes,* mounted 180° apart in the generator frame, rub against the commutator bars. This configuration allows the pair of brushes to contact the commutator bars for one wire loop at a time. And because the armature is turning rapidly, the brushes contact successive commutator pairs rapidly. Each wire loop passes its small voltage charge into a brush. Because there are so many loops, voltage flows from the generator in a practically constant stream.

Generator output is routed directly into the car's electrical system. If there's a heavy load at the moment, current flows through the load. If the load is light, current flows into the battery positive terminal and through its plates, recharging the battery.

But what if there's a low load and the battery is fully charged? What happens to all that current the generator is producing?

If the excess current were not controlled, the generator would continue to push it through the battery, causing the electrolyte to overheat. Excessive gassing would take place inside the battery, boiling the electrolyte out of it, thus ruining it. Obviously, there is a means of controlling excess generator output—

Mechanical voltage regulators are used with DC generators, also with early alternators. One shown at right is adjustable; this one is sealed and must be replaced if it gets out of calibration.

Mechanical voltage regulator controls current output of generator though set of movable points and calibrated coils. Current flows when points are touching. On one shown here, points are adjustable. For best performance, keep points clean. Photo by Ron Sessions.

this is the function of the *mechanical voltage regulator.*

Mechanical Voltage Regulator—As noted earlier, the field coils of the generator become magnets only when current flows through them. What if you could turn off the current to the coils when the battery is fully charged? And what if you could turn field current back on when system voltage starts to drop? That's exactly what the *mechanical voltage regulator* does. Here's how:

Current for the generator's field coils must pass through a set of movable *points* in the voltage regulator. When the points are closed, or touching, current flows. If system voltage gets too high—battery is fully charged—a specially calibrated electromagnet in the voltage regulator becomes strong enough to pull the points open, shutting off current to the coils. So, the field loses its magnetism and the generator stops generating. As soon as system voltage drops below the calibration point of the electromagnet, the points snap shut again, energizing the field once more. This regulation process happens constantly, at a rate of up to 200 times per second. It effectively limits the generator's output to provide only what's needed to power the electrical loads and maintain a full battery charge.

Strictly speaking, a voltage regulator consists only of the electromagnet and

point set just discussed. These are mounted in a metal housing, which also contains a *current regulator*. The entire unit is commonly referred to as the voltage regulator.

The current regulator comes into play when there are many loads working and high generator output is needed. In this case, the voltage regulator stays out of action, permitting maximum generator output. But under these circumstances, it's possible for the generator to work so hard that it burns itself out. Excess current flow is the culprit. To guard against this, the current regulator—a second calibrated electromagnet and point set in the voltage-regulator housing—temporarily cuts off field current if generator *output current flow* is too high.

Most mechanical voltage regulators used in recent years are not adjustable, so there's nothing you can do if the calibration of one of the internal components gets out of specification. Earlier models are adjustable, so consult the appropriate shop manual to see if yours is. AC generators (alternators) use *electronic voltage regulators,* discussed below.

AC GENERATOR (ALTERNATOR)
In the early 1960s, the demands on a car's electrical system had become so great that the DC generator could not handle them. So, the *alternator* was de-

veloped. It is much more efficient than the DC generator, and is able to provide a great deal of current, even at engine idling speed.

Like a DC generator, an alternator relies on the relative movement of a wire loop and a magnet to create current. But there are some major differences.

In an alternator, the *magnet* rotates, while the *wire loops* are stationary. To provide electricity to make the magnet work, an alternator uses carbon brushes, operating against continuous *slip rings*—they're not segmented like a commutator. This means the magnet gets a constant supply of battery voltage, no matter what position it's rotated to. The wire loops are arranged in the generator housing so the rotating magnetic field cuts through each loop in turn.

But, because the rotating magnet has north and south poles like any other magnet, each wire loop gets a shot of "north" followed by a shot of "south" on each revolution. Remember that the flow direction of induced current depends on the direction of the magnetic field or movement. So, each loop gets current induced in one direction, then the other, producing *alternating current.* But auto electrical systems operate on DC, not AC.

The breakthrough that made alternators practical for use in DC electrical systems is an electronic device called a *diode.* It permits current to flow in one direction only. In an alternator, alternating current passes through a set of diodes that change it into a series of one-way current pulses separated by intervals of no current flow. Because these pulses happen so rapidly, the output appears to be a constant stream of DC current.

Electronic Voltage Regulator—Although the alternator originally was teamed with a mechanical voltage regulator, engineers soon found a way to transistorize regulator functions. This led to the development of the *all-electronic voltage regulator.* In time, its size was reduced to the point that it could be installed inside the alternator housing. It has no moving parts and is not serviceable or adjustable in any way. If the voltage regulator becomes defective, you replace it. However, its electronic components are extremely reliable, so replacement isn't often needed.

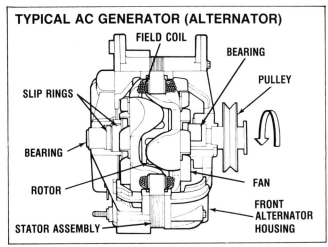

TYPICAL AC GENERATOR (ALTERNATOR)

FIELD COIL

BEARING

PULLEY

SLIP RINGS

BEARING

ROTOR

FAN

FRONT ALTERNATOR HOUSING

STATOR ASSEMBLY

Modern AC generator (alternator) consists of a rotating magnet (field coil) and a stationary coil (stator). Brushes carry current into field coil via slip rings. Drawing courtesy of Chrysler Corp.

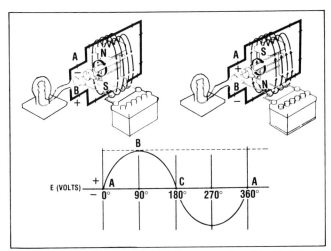

E (VOLTS)

During each revolution of field coil, stator gets a "shot" of South magnetic field and one of North magnetic field. This reversal of fields from moving poles is why alternator creates alternating current. Drawing courtesy of Chrysler Corp.

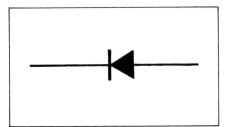

Arrowhead in diode symbol points in direction of current flow. It cannot flow in opposite direction, so diode acts as a one-way electrical valve, transforming AC current to DC current.

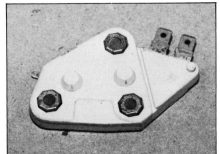

Most modern alternators contain built-in electronic voltage regulator. Entire device is molded into heatproof, weatherproof ceramic and is not serviceable except by replacement. Because of their extreme reliability, this is not often required.

GENERATOR SELECTION

Because a generator is operated by an engine-driven belt, it uses a significant amount of horsepower. And the higher the output rating of the generator, the more horsepower it uses. In a race car or dune buggy, horsepower loss is a serious consideration. However, most competition vehicles have relatively low current demands—ignition and fuel pump, mostly—so they can get by with a low-output generator. On the other hand, an off-road vehicle with a bumper full of driving lamps will require a high-output generator, thus horsepower must be sacrificed.

So, how do you know how big is big

enough? It all boils down to generator output, rated in *amperes*. The amperage output must be *at least* equal to the total current draw of all the electrical loads in the system. Otherwise, the battery must supply current to keep everything running. There should also be enough *excess output* to keep the battery charged. Here are some guidelines:

Restorations—Go with the factory-original part, if authenticity is important. Just be sure generator output is up to specification. As extra insurance, have the old one rebuilt and tested by a generator shop.

Race Cars, Dune Buggies & Kit Cars—Go at least one size larger than

the minimum needed to power all loads in the system. Add up the *current draw* of each electrical device you plan to use, including the ignition coil.

Everyday Vehicles—If you're just adding a pair of driving lights, a stereo, or similar accessory to your street car, the factory-original generator should be adequate. But if you're into competing with the light output of the sun when driving at night, your driving lamps may overtax the standard generator, so it's best to go one size larger. Many auto manufacturers provide a higher-output generator on their cars equipped with air conditioning, and one of these factory generators may be suitable to handle the accessories you're installing.

If you're buying a used or rebuilt generator, or salvaging a one from a parts car, how do you determine its output? First, physical size isn't much of a clue. You must look for an identification code on the generator. It may be a color, a part number or the actual ampere rating. Compare the code to the manufacturer's specifications in the appropriate shop manual or dealer parts book.

Electrical Paths

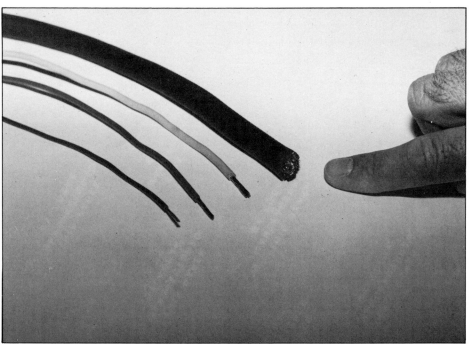

Although wires and cables are both used to conduct electricity, cables are generally much heavier. From bottom to top: 18-gage wire, 14-gage wire, 12-gage wire and battery cable.

Wires are the most obvious electrical paths in an automotive electrical system because they carry current to the various electrical components. The internal circuitry of some modern electrical components uses *printed circuits* instead of wires, so they are another type of electrical path. To complete the circuit between an electrical source (battery) and the load (electrical component) a return path, called a *ground,* must be provided. As previously described, the engine block, metal body and chassis act as the ground between the load and battery negative terminal.

WIRES

Wires can be compared to water pipes, blood vessels or highways. They provide a path for electrical current to flow. The metal wire itself is called a *conductor* because it conducts, or carries, the current. The conductor is covered with *insulation* to keep the current in its correct path—prevent short circuits.

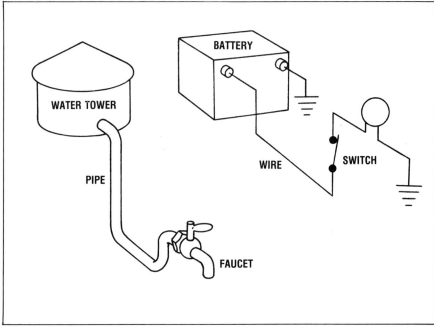

An electrical system can be compared to a water system. Voltage (water pressure) from battery (water tower) provides electricity (water) to flow through wires (pipes) to load (faucet).

WIRE GAGE SELECTION GUIDE

12-Volt System			WIRE GAGE FOR CIRCUIT IN FEET							
Amp	Watts	Candle-power	3'	5'	7'	10'	15'	20'	25'	30'
1	12	6	20	20	20	20	20	20	20	20
1.5	18	10	20	20	20	20	20	20	20	20
2	24	16	20	20	20	20	20	20	20	20
3	36	24	20	20	20	20	20	20	20	20
4	48	30	20	20	20	20	20	20	20	18
5	60	40	20	20	20	20	20	20	18	18
6	72	50	20	20	20	20	20	18	18	16
7	84	60	20	20	20	20	18	18	16	16
8	96	70	20	20	20	18	18	16	16	16
10	120	80	20	20	20	18	18	16	16	14
11	132	90	20	20	20	18	16	16	14	14
12	144	100	20	20	20	18	16	16	14	14
15	180	120	20	20	20	18	16	14	12	12
18	216	140	20	20	18	16	14	14	12	12
20	240	160	20	20	18	16	14	12	12	10
22	264	180	20	18	16	14	12	12	10	10

Typical Electrical Circuits	Approx. Amp.
Air Conditioner	12-20
Back-up Lamps (2)	3-4
Brake Lamps (2)	3.5-4
Cigarette Lighter	10-12
Clock	0.3
Dome Lamp	1
Headlamp Dimmer	2
Headlamps (2)(Low Beam)	8-9
Headlamps (2) (High Beam)	13-15
Heater/Defroster	6-10
Horns (2)	20
Ignition	1.5-4
Instrument-Panel Gauges	0.7-1
Instrument-Panel Lamps	1.5-3
License Lamp	0.5
Parking Lamps (2)	1.5
Power Door Locks	3-5
Power Seat	25-50
Power Windows	20-25
Side-marker Lamps	1.3
Starter Solenoid	10-12
Taillamps (2)	5

Note: Lengths in feet represent total length of wire in circuit, using ground return. In two-wire circuits, be sure to include total length of both wires in circuit. Where 20 gage is listed in chart, 18 gage is recommended for mechanical strength. Gage recommendations are for typical street vehicles, based on a maximum voltage drop of 10%.

A general rule to follow is to use a wire size that gives a voltage drop of 0.5 volt or less across the length of the supply (B+ to load). Here's an example: Suppose you're wiring two driving lamps, each drawing 100 watts, and the supply lead needs to be 18 feet long. Should you use 14 gage wire?

From Ohm's law ($E = I \times R$), you need to know the current draw of both lamps (I), and the resistance of 18 feet of wire (R) in order to calculate the voltage drop across the wire (E).

A derivation of Ohm's law—P (power in watts) = $I \times E$—allows you to calculate the current draw (I) of the two 100-watt lamps:

P (200 watts) ÷ E (13.5 system volts) = I (15 amp).

Next, from the wire size chart at right, you find the resistance of 18 feet of 14-gage wire is 0.0028 ohms per foot.

0.0028 X 18 = 0.05 ohm.

With this information, you now have all the variables of Ohm's Law to solve the problem:

E = I (15 amp) X R (0.05 ohm) = 0.75 volt.

From this, you find that 14-gage wire exceeds the recommended voltage-drop limit of 0.5 volt. So, you go to 12-gage wire, which has a voltage drop of 0.0018 ohms/ft:

E = I (15 amp) X R (0.0018 X 18 ft.) = 0.49 volts.

WIRE SIZES

Metric (mm2)	SAE AWG (gage)	Ohms per 1000 Feet
0.5	20	10.0
0.8	18	6.9
1.0	16	4.7
2.0	14	2.8
3.0	12	1.8
5.0	10	1.1
8.0	8	0.7
13.0	6	0.4
19.0	4	0.3
32.0	2	0.2
40.0	1	0.14
50.0	0	0.11
62.0	00	0.09

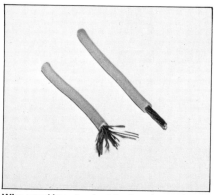

Wire used in auto electrical systems should be stranded to provide maximum flexibility with good conductivity. Example at left shows individual strands that make up conductor. Identical piece at right has not been unraveled.

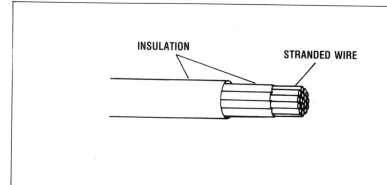

Wire conductors are coated with plastic insulation to prevent short circuits should wire accidentally touch a ground. Drawing courtesy of Chrysler Corp.

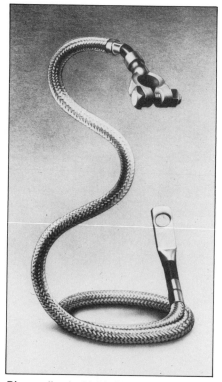

Diamondback shielded battery cable from Taylor Cable Products Inc., features braided-metal alloy shielding over conventional plastic insulation for additional protection and classy appearance. And no, it doesn't bite! Photo courtesy of Taylor Cable Products Inc.

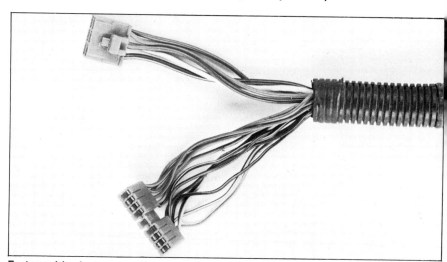

Factory wiring harnesses often use striped wire called *tracer wire*. Adding a red tracer to a white wire, for instance, provides an additional "color" for the harness designer to use. Tracers also help service technicians trace future wiring problems. Photo by Jim Barrett.

Wire comes in many sizes and types. Some wire manufacturers call all their products *wire;* others prefer the term *cable.* By my definition, the main difference between a wire and a cable is *size*— that is, the effective diameter of the conductor. Because a cable must carry a great deal of amperage to operate the starter motor, or a great deal of voltage to operate the sparkplugs, it is relatively large in diameter.

Wires, on the other hand, must carry relatively less amperage, so they have noticeably smaller diameters than cables. Throughout this book, the term cable is used to identify the main conductors used in the starting and secondary-ignition circuits and at the battery. All other conductors in the system are called *wires.*

Sizes—The conductor diameter of a wire or cable is carefully matched to the amperage the wire is expected to carry. It is measured in one of two ways.

Gage is the standard SAE designation for wire sizes, usually referred to as AWG (American Wire Gage). Gages are not measurements of actual dimensions, but are arbitrary numbers that indicate *relative diameters.* And, unlike dimensional measurements, the *larger* the gage number, the *smaller* the wire. Gages used in automotive wiring range from 10 (largest wire diameter) to 20 (smallest wire diameter). Cables range from 00 (largest) to 6 (smallest). The size distinction between wires and cables gets blurred at about gage 8. Gage sizes are in increments of 2, as shown in the chart on page 23.

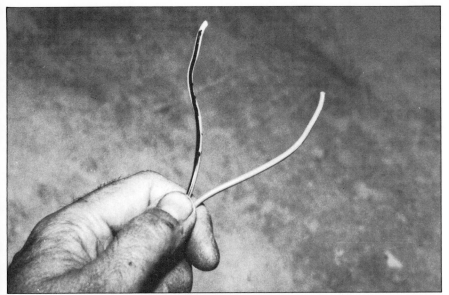
Tracer wire isn't readily available in aftermarket, but you can make tracer wire by drawing stripe on wire with waterproof marking pen.

The second way wire size is specified is by the *area* of the wire's conductor cross-section, expressed in square millimeters (mm^2).

The chart on page 23 compares SAE wire gages with corresponding metric sizes. The chart also shows the relationship between *electrical resistance* and wire size, expressed in ohms per 1000 feet. The larger the wire size, the less the resistance; the less resistance, the greater the maximum possible current flow.

The wire-selection chart on page 23 shows the recommended wire gages to use for circuits of various amperages. The amperage draw of typical automotive circuits are also shown. Use the chart to help you select the correct-gage wire for the various circuits in your car's electrical system.

Conductor Materials—The conductor material used in car wiring is generally copper, although some copper-clad aluminum cable is available. Copper is used for its durability and low electrical resistance.

Conductors come in several strand configurations for various uses. Because of its relatively large cross-section, *single strand wire* is least desirable where flexing is likely to occur, so it is rarely, if ever, used for automotive wiring.

Multiple-strand wire, with its large number of relatively small individual strands, is less likely to fracture from repeated flexing. Generally, for any given overall wire gage, the greater the number of strands, the more flexible the wire. For example, 18-gage wire can contain seven strands of 26-gage wire, 19 strands of 30-gage wire or 41 strands of 34-gage wire. For best flexibility, select wire containing the maximum number of strands for its gage.

Rope-stranded battery cable is often specified where extreme flexibility is needed to facilitate routing.

Braided ground straps, a form of cable, used to be commonly used as battery ground cables. Smaller versions are sometimes used to ground the engine to the chassis, or sheet-metal body to the chassis, if rubber mounts between these assemblies restrict electrical continuity.

Insulation—Regardless of size, all wires and cables have one thing in common—they need an insulative covering to keep the electrons in their correct paths. Years ago, wire and cable insulation was made of braided cloth; today's cars use wires with a more-durable PVC plastic insulation. Several specialty houses carry cloth-insulated wire for antique or collector cars. See the supplier's list on pages 152-153. Whether cloth or plastic, automotive-wire insulation has several important characteristics.

Insulation color is the first characteristic. Each wire of a factory-made wiring harness has its own unique color. This was done so each circuit could be identified as the car was being built, and later for diagnosing a problem. Think of wire colors as street names.

Sometimes, a factory-made harness has more wires in it than there are single wire colors available. In this case, the designer specifies a second color, called a *tracer,* which is a contrasting stripe running the length of the wire. For instance, by adding a white tracer to a red wire, you have distinguished the wire from a plain red one. Tracers are always chosen to contrast with the base color of the insulator, and may come in a number of colors, though purple and orange tracers are rarely seen.

In designing circuit diagrams and building harnesses for your project car, you have complete control over the choice of wire colors. And, because you're going to take extra care in constructing your harness, wire by wire, you probably won't have much need for tracer wire. This is just as well, because tracer wire is not readily available in the aftermarket, anyway.

If you must use tracer wire and you can't find any, you can make your own by using a waterproof marking pen to draw a contrasting stripe along the length of the wire.

Another way to make a tracer, though less desirable, is to mark the terminal ends of the wires with a dab of paint. There are several disadvantages to this method. First, the paint may chip off. Second, there's no way to distinguish the tracer wire except at the end. Third, if the wire turns out to be too long, and you cut it short, the tracer may end up on the garage floor.

Finally, labels can be purchased at many electrical-supply outlets, for the sole purpose of marking wire ends for identification purposes. See page 142.

One more thing about color: On American-built cars, *black wires* are almost always used on the ground (−) side of a circuit. However, in the thousands of factory harnesses made, there are probably some exceptions to this rule. German cars use *brown*—the color of earth (ground)—to indicate ground.

Red always means hot (+) although other colors are also used for hot wires in a harness. Imagine the confusion if every feed wire in the harness was the same color!

The red-for-positive convention also applies to the battery positive cable and battery-feed wires to the generator, fuse panel and ignition switch. In most cars, these are the wires that are always hot, even when the ignition switch is off. Again, you may find exceptions.

Another important characteristic of a wire's insulation is its *durability*. If you were buying wire by the truckload like an auto factory does, you could order wire with insulation to fit your specifications. But most of us buy wire by the spool at local auto-supply stores. To be sure you're getting wire with insulation best suited for most common vehicle applications, stick with name brands. Packard Electric, a division of General Motors, is perhaps the most widely known wire manufacturer. Another brand widely available in parts stores is distributed by JT & T.

Depending on application, the *temperature rating* of wire insulation can be important. Wire for common applications has insulation rated at 85C (185F).

This rating is adequate for most common applications. However, if you're building a race car, use wire with an insulation rating of at least 105C (221F). For more on high-temperature wire insulation for race cars, see page 140.

Whatever brand you buy, don't skimp on quality to save a few dollars. In the long run, it's false economy if the insulation melts from under-hood heat or dissolves in grease and causes a short circuit. Burned wires or a car fire is a high price to pay for bargain-basement materials.

PRINTED CIRCUITS

Wires and cables aren't the only conductors found in a vehicle's wiring system. On newer cars, some factory-installed components contain printed circuits (PCs) that conduct electricity just like wires do. You'll find printed circuits most often in original-equipment instrument clusters because they provide a savings in labor cost for the factory.

There was a time when each instrument and light bulb in the instrument cluster was wired separately. This, incidentally, is the way you'll wire individual or add-on instruments in your project car.

Nowadays, car manufacturers are looking for ways to reduce hand labor and its related costs, to keep car prices under control. The one-piece printed circuit does this nicely. And its nearly error-proof design provides higher quality in mass-produced cars.

A printed circuit consists of metal foil current paths adhered to a flexible plastic backing (PC board). Each lamp socket and instrument connector plugs into the circuit at the appropriate place. A cavity is provided for plugging in the wiring-harness connector. As long as the harness is correctly made, circuit continuity is virtually assured.

Chances are, you won't run into a printed circuit in restoring a collectible or antique car, or if you're installing individual instruments, as mentioned earlier. But, if you're installing an entire *factory-made instrument cluster* in your project car, there are several important things to consider when dealing with printed circuits.

First, the harness you make for your car must be identical to the one used in the car the cluster was originally designed for. The best thing to do here is to cut the printed-circuit connector from the harness in a junked car, leaving about 1 foot of wire extending from each cavity. Then, when you build your own harness, connect the wiring directly to the cut-off wires, following the factory's orientation diagram for each circuit. If this isn't entirely clear, it will be when you read the chapters on circuit design and harness construction.

Treat a printed circuit gently. It's rugged enough to withstand years of service inside a car's instrument panel, but it won't hold up to manhandling. Once a foil circuit is scraped off or broken, there's no practical way to repair it. So, take care when removing or installing the harness connector—be firm and deliberate, but don't use a twisting motion. When installing lamp sockets, be sure the foil is not damaged by the socket tabs.

To illustrate this point, I recently had to replace the printed circuit in one of the family cars. The instrument-cluster illumination and tachometer suddenly and simultaneously quit working. The problem turned out to be a circuit-ground foil that was damaged when the car was built. It held up for several years, but eventually either road vibration or normal expansion and contraction caused it to fracture. Problems like this can be particularly frustrating to trace because

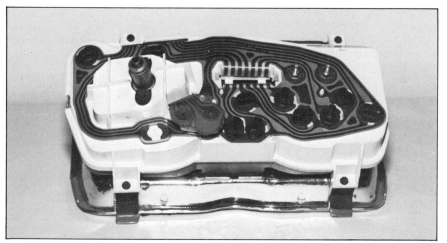

Factory instrument clusters often use printed circuitry. Foil paths are adhered to flexible plastic carrier. Each electrical component in cluster has a terminal that contacts the appropriate foil to complete the circuit.

Instrument-panel harness is connected to printed circuit with a push-in connector. Sides of connector contact foils on printed circuit. If you plan to use an OEM printed circuit in your car, cut attached wires at least 1 foot from terminal, if possible, to allow plenty of splicing room.

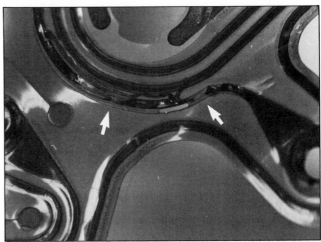

Printed circuits don't handle excess current flow too well. In fact, they tend to act as a fuse. Damage like this (arrows) is fairly easy to locate but impossible to repair. Failure due to fracture in foil circuit is much more difficult to locate, and equally impossible to repair.

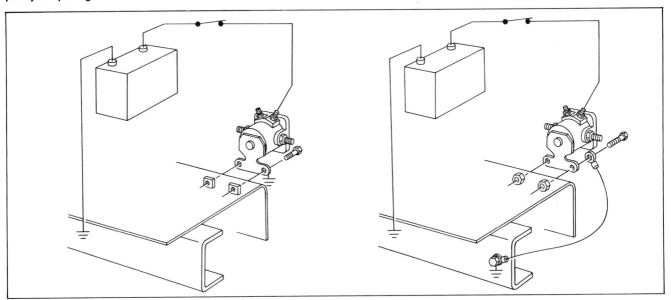

Although easy to overlook, ground circuits are necessary part of electrical path. Most components are self-grounded through their mounting screws to metal body, frame or engine (left). Such components won't ground through fiberglass body because it acts as an insulator. You'll need to run separate ground wire as shown at right.

the break was hidden under the edge of one of the lamp sockets.

GROUNDS

Almost every circuit has one current path that's easy to overlook, even though it's the biggest. It is the car's *entire structure*—chassis, engine and metal body. Together, they form a *common return path* or ground. They're all grounded to the battery's negative terminal. Many electrical components ground directly through one of these, simply by being bolted in place. This type of electrical path is sometimes called a *self-ground*, meaning the component doesn't need a separate (floating) ground wire.

In modern, lightweight cars with numerous plastic body components, as well as in fiberglass-bodied cars, non-conductive body parts won't carry current, so separate ground wires are needed for self-grounded components attached to these. The ground wires are attached to the metal chassis at a convenient point. Often, ground wires for several components connect to a single *ground terminal* or *stud*.

If the component you're installing was designed to have a self-ground, you must provide a means of attaching the ground wire to the component. Often a *ring terminal*, page 63, can be used on the component's mounting stud.

Electrical Controls

Toggles are probably most commonly used switches in race cars of all types. Those being installed on Chevy S-10 off-road racing pickup truck are not fancy, but they're high-quality, reliable switches that are easy to operate under racing conditions. Photo by Tom Monroe.

The driver's interaction with most mechanical systems in a car is through some sort of *control device*. And usually, the device being controlled performs the same way each time the control is operated. The steering wheel is one example. Although the end effect is to cause the front wheels of the car to change their angle to some degree or other, the driver's input is always the same—he turns the steering wheel, a little or a lot, and the front wheels respond accordingly.

In an automotive electrical system, control devices can be placed in three main groups—*switches, solenoids and relays,* and *rheostats.*

Switches are used to turn circuits on or off. Most switches are intended to be operated directly and consciously by the vehicle's occupants—the headlamp or radio switch, for instance. Others are operated by the occupants only as the consequence of performing some other action. Door-jamb dome-lamp switches are good examples. The switches are op-

erated indirectly by opening and closing a car door.

Still other switches are operated by *conditions* in a car and may or may not come under human influence. For example, the park switch inside a windshield-wiper motor is operated by a cam inside the motor housing.

Solenoids and relays automatically activate electrical devices or secondary circuits, either as a result of some action taken by the driver or by changing conditions in the car. For instance, the solenoid for the starter motor is activated when the ignition switch is turned on to start the car.

Rheostats control the amount of resistance in a circuit, thereby controlling the amount of current fed to an electrical component. The dimmer control for instrument-panel illumination is an example.

The type of electrical control operated directly by the driver is almost always a switch or rheostat, although the part of the control actually being operated may

take the form of a knob, lever, handle or push button.

Often, the driver operates the control of a mechanical or electrical system because he's sensed the need to do so from something he's observed. The action could be in response to something directly observed in the environment— such as operating the horn to warn a child darting into the road. Or, it could be a result of observing a condition by means of a *display.* For example, you dim the headlamps for an approaching car because you notice the headlamp high-beam indicator is illuminated. Because displays are designed to convey information, their design and location are also influenced by human factors. Displays are discussed on pages 38-48.

In the following discussion on controls, a distinction is made between *original equipment manufacturer* (OEM) controls that are installed in a car at the factory, and *aftermarket* controls that you can buy at auto-parts stores or through mail-order parts catalogs.

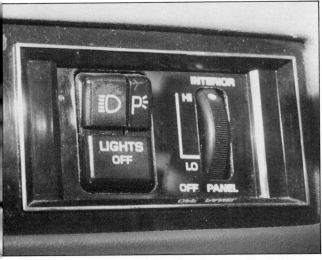

In recent years, car makers have begun using rocker switches like ones shown at left. These switches provide a positive feel when operated, are easy to identify by touch, and look good. Dial at right is rheostat control for dash panel illumination.

OEM switches are often difficult to adapt to a hobby car. For example, this heater control is built into a plastic bezel that would be hard to relocate without a lot of rework.

Good example of how OEM switches can be integrated. Each control is easily identified by touch. Headlamp switch is conventional push-pull type, while attached fog lamp switch is operated by a rotating thumb tab (left). Small slider switch for electric mirrors (right) selects right side or left, depending on which way it is moved. Mounted directly above it, "joy stick" is used to control mirror-head movement.

Windshield-wiper switch mounted on steering column of late-model GM car would be all but impossible to adapt to location other than original. When you encounter such a situation, you'll have to use your ingenuity to adapt aftermarket controls.

Most OEM headlamp switches contain built-in circuit breaker for headlamp circuit. Park and taillamps, although controlled by same switch, usually rely on separate fuse.

OEM controls are almost always designed for specific jobs and fit into specially designed mounts and bezels. Because of their single-purpose design, they don't lend themselves to kit cars and custom applications as easily as the aftermarket controls do.

SWITCHES

In its simplest form, a switch is nothing but an on/off device capable of closing and opening a circuit. Switches come in a wide variety of configurations. Some are seemingly complex in operation but their basic function is to turn equipment on and off.

Most switches are *normally open,* meaning when they're *relaxed,* the circuit is open, or disconnected. In fact, some switches are spring-loaded to return automatically to the OFF position when you remove your finger from the switch. These are called *momentary-on* or *momentary-contact* switches. Most momentary-on switches are operated by a push button.

Some switches are made with built-in *circuit protection* in the form of a circuit breaker or fuse. Aftermarket switches are more likely than OEM switches to have this kind of protection, because the aftermarket switch is often used to control an accessory circuit added to the car by the owner.

The built-in fuse or breaker does away with the need to find a vacant slot in the

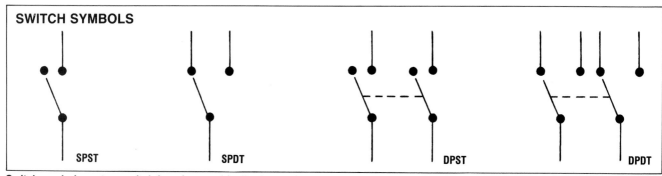

SWITCH SYMBOLS

SPST   SPDT   DPST   DPDT

Switch symbols portray switch function graphically. From left to right: *Single-pole, single-throw (SPST)* switch is used to control a single circuit. *Single-pole, double-throw (SPDT)* switch can select between two different circuits, and will operate only the one selected. *Double-pole, single-throw (DPST)* switch can operate two entirely separate circuits with one control. Both circuits are either ON or OFF at the same time. *Double-pole, double-throw (DPDT)* switch can select between four different circuits, two at a time. It is essentially two DPST switches in a single housing.

car's fuse panel. However, if the switch you're using has no built-in circuit protection and there's no place in the fuse panel to add another, use an *in-line fuse*. This is simply a fuse and fuse holder wired into the circuit between the load and the power source.

OEM switches usually don't have built-in circuit protection because most original circuits route through the car's fuse panel. One notable exception, particularly in American-built cars, is the headlamp switch—instead of using a fuse in the fuse panel, the switch contains a self-resetting circuit breaker. Circuit breakers are covered in depth on pages 35-36.

Switch Types—Switches can be categorized by their *electrical function,* or how the switch actually works. Within each category, you may find several different kinds of *switch actuators,* described later.

Single-pole, single-throw is the simplest kind of switch, with only two positions—ON and OFF. It's often referred to in parts catalogs by the initials *SPST.* Contrary to a popular misconception, the *pole* does not refer to the actuator, but the movable element inside the switch that makes and breaks the circuit. In most switches, you can't even see the pole unless you remove the housing.

Push-button-operated SPST switches turn the circuit on when pushed, and off when released. This is the most common form of momentary-on switch, which is the kind of switch you'd use on your house for the doorbell. Race-car starter

buttons are another example. Also, a switch that activates a windshield-washer pump is of this type. Another type of push-button switch turns the circuit on with the first push and off with a second push. You'll occasionally find rotary SPST switches operated by a knob, but they're not common in cars.

Regardless of the type of control, the SPST switch has one characteristic that distinguishes it from all other switches. Because it has only one pole and only two positions—ON and OFF—it can control only one circuit. The SPST switch has two terminals, one connected to the battery and one to the load.

Single-pole, double-throw is the next most-complicated switch. Its abbreviated name is *SPDT.* This switch also has only one movable element inside, but it has three terminals outside. One terminal connects to the battery. The other two connect to two separate circuits or loads. This means one load or the other can be connected to the battery, depending on the position of the switch actuator. Some SPDT switches have three actuator positions—ON, OFF, ON. This is called a *center-off* switch. It controls one load in each of the ON positions, and breaks both circuits when the actuator is in between. Some SPDT switches are the momentary-on type.

Double-pole, single-throw switches are next in complexity. The abbreviation is *DPST.* This switch has only two positions—ON and OFF. But unlike the SPST switch, it has *two independent poles,* each connected to its own circuit.

There are four terminals, two for each pole. The DPST is really two SPST switches in a common housing, with a common control actuator. It is used to operate two separate loads simultaneously.

Double-pole, double-throw switches are the most complex of the common aftermarket switches. OEM switches, on the other hand, can become almost grotesquely complex in function, but that is another matter. The DPST switch has two independent poles, each with two ON positions. Actually, it's simply two separate SPDT switches in one housing, electrically independent of each other, operated by one control actuator. The DPDT switch has six terminals, three for each pole. DPDT switches come in two designs—ON, OFF, ON and also MOMENTARY ON, OFF, MOMENTARY ON.

All the switch types mentioned above—no matter how complex—serve only to turn circuits on or off. Another type of switch, called a *single-pole, multiple-throw switch,* allows you to select various settings on multiple-speed devices, such as a heater fan. The switch is wired to a series of resistors that control the current being fed to the device. Typical settings are: OFF, LOW, MED, HIGH.

Switch Actuators—Although switches can be categorized by electrical function, they are usually identified by *actuator* type. The actuator is the part you touch to operate the switch, such as a knob, toggle or push button.

Toggle switches are readily available in aftermarket and come with a wide variety of actuators (control handles). When selecting a toggle switch, be sure it has correct combinations of poles and throws for circuits it will be operating.

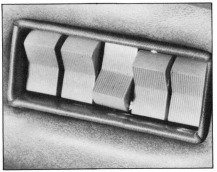

Rocker switch is actually a toggle switch with a unique actuator. They look more "finished" than a row of toggle switches, though more difficult to install. They also reduce the likelihood of actuating switch if you accidentally bump into it.

Push-button switches in cars are usually the *momentary-on* type. Examples include horn button, windshield-washer button, and starter button on older cars and some race cars. Switch shown here is ON/OFF type. Push it once for ON, once more for OFF.

Toggle switches have a tapered handle, usually operated by flipping it with your fingers. Some manufacturers call this a *bat switch* because the actuator resembles a baseball bat in shape. The operation of a toggle switch should offer *elastic resistance;* that is, resistance that builds, then decreases as the contact position is approached.

Rocker switches are a more stylized version of the toggle switch. The actuator rocks like a playground teeter-totter. You operate the rocker by pushing one end or the other with your fingertip. Like the toggle switch, the operation of these switches should offer elastic resistance.

Push-button switches are self-explanatory. A button, often round, protrudes from the switch body. It's operated by depressing the button with your fingertip. Push buttons should have a concave surface for best finger contact. ON/OFF types should offer an audible, feelable click as the button is pushed. Resistance should build with increasing pressure, and suddenly release as the switch operates.

Rotary switches have a knob that is designed to be turned. On some switches, the knob has a pointer to indicate switch position. Rotary switches are not often used in automobile systems, except possibly to control a heater fan. Rotary switches are far more common in electronic devices such as radios. It is desirable that a rotary switch offer a discernable detent at each selector position, with resistance decreasing as the detent is approached. Minimum distance between

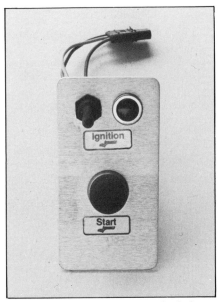

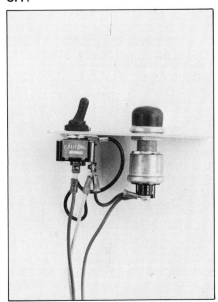

Race car ignition switch setup includes toggle ON/OFF switch with indicator lamp and *momentary-on* push-button start switch. Actuators are protected by rubber "boots" to reduce dirt and dust infiltration. Side view shows mounting and wiring details. Prewired and premounted to brushed chrome panel, this setup is available from Longacre Automotive Racing Products. Photos by Jim Barrett.

detents should be 1/4 inch.

The following chart compares the desirability of various switch actuators against operating characteristics.

RHEOSTATS

One of the technical terms for a rheostat is a *variable resistor*. This term is appropriate because it indicates how the rheostat works. Rheostats are also referred to as *potentiometers*.

A piece of resistance wire, usually in coil form, is connected at one end to a battery feed. The other end of the resistance wire isn't attached to anything. The wire going from the rheostat to the load is attached to a movable element called a *wiper*. The wiper is arranged so it rubs against the length of the resistance wire. When the wiper is close to the battery-feed end of the resistance wire, the rheostat adds little extra resistance to the circuit. As you slide the wiper to the far end of the resistance wire, current flows through increasingly more resistance to get to the wiper, reducing current flow to the load.

Within the design range of the rheo-

Coil of resistance wire in headlamp switch is a *rheostat,* used to control intensity of gauge illumination. Because resistance wire heats up during operation, it is installed in heatproof ceramic housing.

Switches mounted on backside of OEM air conditioner grille. Rheostat at top is used to control A/C thermostat (note numbered dial); bottom switch is a single-pole, multiple-throw switch to control fan speeds. Photo by Jim Barrett.

stat, you can dial in no resistance up to high resistance, or any amount in between. This gives you precise control over the amount of current being fed to the load. The most common use for a rheostat in a car is the dimmer for the instrument-panel illumination lamps. In most OEM headlamp switches, the dimmer rheostat is operated by twisting the headlamp knob.

Rheostats designed for automotive applications are usually not capable of carrying high current without burning out. Circuits with high currents in need of variable resistance are usually controlled by a single-pole, multiple-throw slide switch. The selector for a three-speed heater-blower motor fits this category. The selector switch regulates motor speed by selecting various resistors in series with the motor. In the LOW position, the full set of resistors is selected. Each higher speed bypasses one more resistor, until, in the HIGH position, all resistors are bypassed, providing full current to the motor. The resistors are capable of carrying high current without burning out. They do get hot, though, so they're usually located in the blower airstream.

GUIDELINES FOR SELECTING CONTROLS

Regardless of the nature of the control device, you must consider several objectives when selecting one.

First, the control should be *capable of performing its function effectively* when it is operated as it is designed to. Second, it should be *suitable for use* in terms of feel and the limitations of the human hand. These two objectives are usually met by a single design. If it fits your hand, it probably will function the way it was intended.

For mass-production cars, the designer must take into account the entire range of probable drivers, from the smallest to the largest—or in some cases, for a compromise between these extremes. When you're selecting the controls and their locations for your project car, you're dealing with your own physical characteristics, and possibly those of a few of your family members. So your choices can be more selective.

When selecting a control for a particular electrical component, it must be electrically compatible with the device it's going to operate. Beyond that, there are several other factors to consider, listed in the chart below.

In the chart, *discrete* means being able to select and hold a specific setting. For example, a simple ON/OFF switch is capable of selecting and holding two specific settings. *Quantitative and continuous* mean being able to select a variable setting. An example is the rheostat that controls the brightness of instrument-panel illumination. Another example is the multiple-throw switch that selects various heater-fan speeds.

Another consideration in selecting a control is the way you're able to identify it. For instance, in the chart below, shape and texture is listed as an identifier. This means you can identify a specific control by feeling its shape and texture. But, as

SWITCH IDENTIFIERS

The term identifier refers to how you identify a switch.

Identifier	Characteristic
Shape and Texture	Good for low illumination. Limited by number of textures and shapes available. Hard to detect with gloves on.
Location	Good for low illumination. Limited by space requirements.
Color	Good visual identification, but must be viewed directly; requires illumination.
Label	Good visual identification if large. Must be viewed directly. Requires illumination.
Operating Method	Good because switch can usually be operated only one way. Must be tried before identification is certain.

SWITCH ACTUATORS

Actuator Type	Function
Push button	Activation
Toggle switch	Activation, discrete setting
Rotary selector	Discrete setting
Knob	Discrete, quantitative, continuous setting
Lever	Quantitative, continuous setting

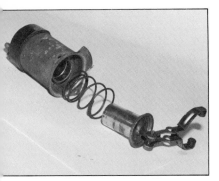

I took this GM starter solenoid apart so you could see the movable core clearly. Core is attached to forked lever to move drive gear when starter is operated.

Newer cars rely on relays, such as this one, to operate computer-controlled emission devices. Computer's low-current circuitry operates low-current relay, which in turn operates high-current device.

Typical Ford starter solenoid. Although it acts as a relay in the battery-to-starter cable, it contains a movable core, so is correctly called a solenoid.

you can see in the Characteristics column of the chart, it's hard to discern shape or texture if you're wearing gloves.

When designing the instrument panel on your car, make sure the type of switch actuators you choose are compatible with their location on the panel. For example, if a headlamp switch and windshield-wiper switch have the same shape and operate the same way, they should not be positioned next to each other. Otherwise, you might turn off the headlamps instead of turning on the wipers.

A final consideration is the way the switch mounts to the instrument panel. Make sure the mounting method is compatible with panel thickness and with the clearance available behind the panel for installation.

So before you buy your switches, refer to pages 76-77 to make sure the switches are suitable for their positions and can be installed without difficulty.

SOLENOIDS & RELAYS

Two definitions are in order because many people incorrectly use the term *solenoid* interchangeably with the term *relay*. About all they have in common is that both rely on an electromagnetic coil for operation.

A *solenoid* is an electromechanical de-

vice consisting of a electromagnetic coil surrounding a movable metal *core* or *plunger*. When the coil is energized, the magnetic field pulls the core until it's centered in the coil. Usually, there's a return spring to pull the core back to the uncentered position when the coil is shut off. The movable core can be connected

to various mechanisms that move each time the magnetic coil is energized.

A *relay* is an electromechanical device consisting of an electromagnetic coil, a fixed core and a movable *armature*. The armature, attracted by magnetism in the core, moves whenever the coil is energized.

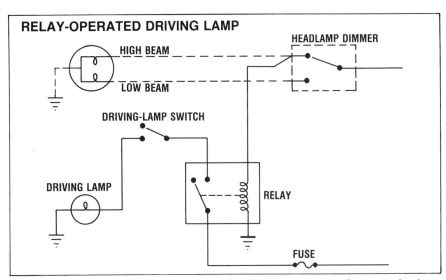

Driving lamp circuits often use a relay. Although there are several ways to hook up circuitry, this hookup permits driving lamp to operate only when high beams are on. In some states, this is required by law. You could wire a fog lamp similarly, using the low-beam circuit or parking lamp circuit to operate relay. An advantage of this system is that you cannot accidentally leave high-current accessory lamp turned on.

33

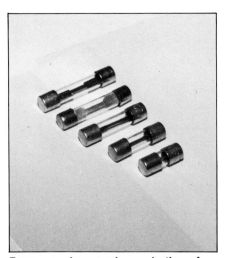

Fuses come in many sizes and ratings. Amp ratings of glass-capsule fuses are often related to physical size—the larger the fuse the higher the rating—though this is not a hard-and-fast rule. Always check rating stamped on fuse.

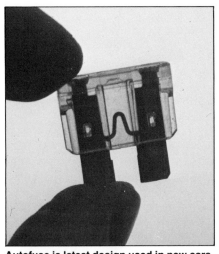

Autofuse is latest design used in new cars. Fuse element is readily visible through transparent plastic body. Two metal blades plug into matching sockets in fuse panel.

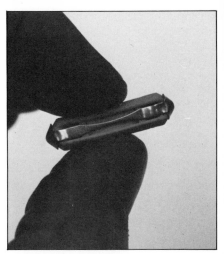

German cars use ceramic-body fuses. Exposed fuse element is recessed into shallow groove on body. Pointed metal cap on each end makes contact with fuse panel.

The distinguishing feature, then, is the core. In a solenoid, the core moves to do the work. In a relay, the core remains stationary and the armature moves.

Solenoid & Relay Applications—The mechanical aspects of solenoids and relays largely determine how they're used. Solenoids are most often found in starter systems. On GM starters, for instance, there's a large solenoid on top of the starter motor. It moves the starter-drive mechanism into engagement when the starter is actuated.

In Ford starter systems, the solenoid is mounted to the inner fender or firewall. It serves as an electrical switch between the battery and the starter motor. When the ignition is turned to the START position, a small current energizes the solenoid coil. This pulls the movable core, which in turn operates the heavy-duty switch controlling starter-motor current.

In both starter-system examples, a solenoid is used because the application calls for a large amount of motion and a large amount of force. Solenoids are excellent for this purpose.

Relays, on the other hand, are usually used in more precise applications. The motion of the movable armature can be made to operate delicate devices, such a set of electrical contacts. Relays are often used to control secondary electrical circuits.

For instance, you're installing a high-power driving lamp, and you want it to come on with the headlamp high beams only. But you feel the headlamp circuit in your car won't carry much more additional amperage. The answer is a relay. Wire it so the coil is energized by the car's high-beam circuit. The coil won't draw much current, so the amperage of the headlamp circuit isn't increased much. Run a feed wire from the battery to one of the relay's armature terminals, then wire the driving-lamp switch and lamp to the other armature terminal. To protect this circuit, wire in a fusible link or in-line fuse. The relay manufacturer provides the hookup instructions. If you lose them, refer to the section on testing, page 127, to determine continuity through a relay.

Here's how your new circuit will work: If you turn on the driving-lamp switch, nothing happens. Why? With the high beams off, the relay coil is relaxed, so the armature passes no current and the driving lamp stays off. But when the high beams are turned on, the relay coil is energized, allowing the armature to conduct battery voltage directly to the driving lamp. The lamp lights without drawing excess current from the headlamp circuit. An added benefit of this setup is that most states require the driving lamp to turn off when the headlamps are

dimmed. The relay insures that you'll never forget.

CIRCUIT PROTECTION

If you could be sure every circuit and electrical component in your car would work without problems for the life of the car, you wouldn't need circuit-protection devices. But you can't be sure, so circuits need some type of device to prevent them from overloading, which can damage both wiring and electrical components. The three basic circuit-protection devices used in cars are the *fuse,* the *circuit breaker* and the *fusible link.*

Fuses—These are the most common form of circuit protection in automotive electrical systems. You can partially deduce their function from the definition of the word *fuse*—to melt or become liquid from the application of heat. Excessive amperage in a circuit can cause a conductor to get hot. Even wire can eventually get hot enough to melt, particularly when a short circuit occurs.

In designing a circuit, you must select wire that will safely carry the expected (normal) amperage plus a safety margin. If this margin is exceeded, the wire will overheat and eventually melt or ignite surrounding items—that is, unless you install the correctly rated fuse in the circuit, in series with the load. When amperage is exceeded, the fuse melts before the wire or electrical device is damaged.

Fuses come in various configurations (shapes) and in various *ratings*. The shape of a fuse has little to do with its effectiveness. Rather, fuse shapes are determined by the manufacturer. The rating of a fuse is the maximum number of amperes it will flow without melting. In the United States, there are two styles of fuses in widespread use.

The *glass-capsule fuse* has existed, largely unchanged, for decades. Generally, you'll find a metal cap on each end of a short glass capsule, or tube. A strip of special metal joins the two caps and is visible through the glass. The metal strip is the actual fuse material—it's the part that melts when current flow exceeds the fuse rating. The glass protects the sometimes-delicate metal strip from mechanical damage and insulates it from accidental short circuits, while providing ready visual access when you need to inspect for a burned-out (blown) fuse.

Drawbacks of the glass-capsule fuse include:
● Fuse length is often related to its amperage rating. This means you may not be able to use a different amperage rating when you add or remove a load from an existing circuit.
● A special tool is needed to remove glass fuses from the fuse panel without breaking them.
● It's possible for the metal strip to blow inside the metal cap, where its condition is not readily visible.
● Fuse ratings are embossed in the metal end cap or etched on the glass capsule. In either case, it's often difficult to read the marking without removing the fuse.

These disadvantages are partially offset by several advantages:
● Because they've been in use for a long time, glass-capsule fuses are inexpensive and readily available.
● The fuse's condition is usually—but not always—obvious when visually inspected.
● Aftermarket fuse blocks are readily available for custom applications. So are in-line fuse holders.

The *Autofuse* was first used in some GM cars in the mid-'70s and is now used on all new GM products. It's finding its way into new vehicles from other manufacturers, as well. These fuses are readily identified by their shape.

The rectangular fuse body is transparent plastic with two metal blade terminals protruding from the back. The top of the fuse is molded to provide an easy finger grip. Two recesses in the top provide a convenient test point, even with the fuse installed. Like the glass fuse, the Autofuse can be visually inspected by viewing the fuse element through the fuse body.

Some advantages to consider:
● Easy to install and remove without tools.
● Easy to visually inspect. Test points are also built in.
● Fuse-amperage ratings are boldly printed on the end, and can be easily read.
● Can be carried or stored without fear of breakage.

These advantages are partially offset by several disadvantages:
● Must be pulled from the fuse panel for visual inspection.
● Cost. Because these fuses are relatively new to the market, their prices are not yet competitive with glass fuses.
● Availability is improving, but glass fuses are more widely available.
● Require a special fuse block, just becoming available in the aftermarket. The best source is still a salvage yard.

German cars use a variation of the glass-capsule fuse. It has a body of solid ceramic, with the fuse element fitted into a groove on the outside. The element is formed at both ends into cone-shaped metal caps that fit into spring clips in the fuse block. Although the fuse element is readily visible, it's vulnerable to damage. And, the end caps are prone to oxidation, which can seriously hamper current flow. Incidentally, the ceramic-type fuse is color coded for easy reading of amperage rating. Volkswagens typically use 8-amp (white) and 16-amp (red) fuses.

Circuit Breakers—A circuit breaker is a more-sophisticated protection device than a fuse. It uses a thermally sensitive element that snaps open when overheated by excessive amperage. But after a short cooling-down period, the circuit breaker resets itself—some types must be reset manually—and begins passing current

Plug-in circuit breaker (left) fits same panel as Autofuse (right). Breaker can be substituted for like-rated fuse, provided you understand consequences explained in text. I successfully used one of these breakers in an electric motor circuit that drew just enough excess current on start-up to blow a conventional fuse. DO NOT use a breaker rated higher than the fuse, however.

again. There is nothing to replace. So, if cost is no object, consider using circuit breakers to protect the wiring in your car. I found a listing in the JC Whitney catalog for circuit breakers ranging from 5 to 50 amperes, but at a price comparable to about 10 glass-capsule fuses.

If excellence is your style, use them. In addition to their durability and reusability, the impressive "high-tech" appearance of a row of circuit breakers may make them worth the extra cost. In the chapter on race-car wiring, I explain how and why race-car builders use circuit breakers instead of fuses.

Circuit breakers of the type just discussed are housed in small metal cans about the size of your first thumb joint. There are two threaded terminal studs, insulated from each other. The studs are used to attach the circuit breaker to the wiring, using ring terminals. *Don't use these studs for mounting the circuit breaker!* A circuit breaker designed for panel-mounting has a flange for that purpose. If there's no flange, the wiring will support it. Personally, I think the unmounted kind appear somewhat crude. Also, you must install them so their terminals don't accidentally contact metal parts that could ground the circuit, causing a short.

A second type of circuit breaker is the fuse-clip type, designed to snap into a standard glass-capsule fuse block. The clip type has the reset advantage of the stud type and carries a similar price tag. But don't get carried away with a wholesale change from glass-capsule fuses to clip-type circuit breakers until you've tried them. Make sure there's enough room for them to fit side-by-side in your particular fuse block.

There's a third type of circuit breaker used on many cars. These are built into the components they protect. For instance, most headlamp switches used in domestic cars contain a circuit breaker for the headlamp circuit only. Other components controlled by the headlamp switch rely on fuses in their separate circuits. True to form, this built-in circuit breaker resets itself after cooling down. Headlamps that slowly flash on and off indicate a short in the circuit, and that the circuit breaker is doing its job.

Some electrical-accessory motors also have built-in circuit breakers, but their purpose is twofold. First, they protect the circuit from normal electrical overloads. But they also break the circuit if the motor overheats dangerously. For instance, this could occur if a windshield-wiper blade froze to the glass.

Built-in circuit breakers usually go through life without need of attention. But when you're trying to diagnose a perplexing operating problem, be aware that the circuit may include one of these breakers.

An Argument Against Circuit Breakers—There's some controversy among electrical experts whether fuses or circuit breakers are better. Those favoring fuses mention several drawbacks of circuit breakers. One is the higher cost. A more important one, though, is the automatic-reset function on many circuit breakers, which appears at first to be a strong positive feature.

Opponents of circuit breakers rightfully point out that after an automatic-reset circuit breaker cools, it permits current to flow again. If the cause of the overload is still there—and it probably will be—the circuit breaker again heats up and breaks the circuit. This *on-off-on-off* cycle repeats until the problem is corrected. Unfortunately, during the *on* time, current flows unimpeded. If the circuit breaker's trip point is too high, the *on* periods may be long enough to overheat the wiring, just as though there were no circuit pro-

tection. Opponents of circuit breakers consider this a major safety hazard. They may be right.

The use of circuit breakers in place of fuses calls for some common sense. If you notice an on-off cycle occurring, immediately shut down the affected circuit, then find and correct the problem before using the circuit again.

The idea that circuit breakers pose a major safety hazard can be countered by noting that almost all OEM headlamp circuits are protected by an auto-reset circuit breaker in the headlamp switch. The theory here is that it's better to have flickering headlamps than to suddenly have none.

Then, there are special considerations. For instance, the wiring and circuit breaker protecting a race-car fuel pump should be "sized" so they will handle more current than the pump! The reason is simple: It's better to finish a race and possibly win thousands of dollars than to save a $50 electric fuel pump.

It all boils down to choice. If you think you're likely to ignore the intermittent warnings of auto-reset circuit breakers, use fuses and carry sufficient spares. Or, in some circuits, you may be able to use the type of circuit breaker that resets manually by means of a toggle switch. These are the type used on race cars, and are readily available. On the other hand, if you've gained enough experience to wire a car from scratch, you can probably deal with auto-reset circuit breakers and benefit from the unique protection they offer.

Fusible Links—If you're building an electrical system from scratch, you may not even be aware of fusible links. But if

---

CURE FOR CORRODED FUSE FAILURES

Fuses used in German cars, as typified by the VW, are prone to corrosion. Normally, electrical contact is made between the conical fuse end caps and the round holes in the fuse-holder clips. Because the contact area is so small, it takes only a little corrosion buildup here to completely open a circuit. Buildup may occur over a period of time, gradually impeding current flow until it finally stops. The symptom is the same as with a blown fuse.

You can remedy the condition by removing the fuse and cleaning it with a suitable abrasive cloth or electrical-contact spray cleaner.

Here's a tip that lets you know in advance that corrosion buildup has started, so you can take preventive action before a full failure happens.

Solder a general-purpose *light-emitting diode* (LED) across each fuse holder. That way, when corrosion buildup starts to resist current flow through the fuse-holder clips, part of the current shunts—bypasses the fuse—through the LED, causing it to light. As corrosion gets worse, the LED glows brighter as more current passes through it. In case of a blown fuse, the LED would glow at full brilliance. This means you could also use this method to indicate blown fuses in a conventional fuse panel, as well. It also gives your car a "high-tech" look.

For correct functioning, the LED requires a resistor wired in series. A 1,000-ohm, 1/4-watt resistor is used with a 12-volt system. For 6-volt systems, use a 680-ohm resistor.

Begin by noting the positions of all the fuses, then remove them. Solder a resistor to one clip at each fuse position. Turn on the ignition key, as well as any other switches that feed the fuses. Temporarily hold the two leads of an LED against the first resistor and the corresponding free spring clip. If the LED doesn't light, reverse the leads. When the LED lights, carefully solder it in place. Be sure all surfaces are clean and bright. *Use only rosin-core solder specified for electronic work.* Avoid prolonged or excessive heat. I prefer a small pin-tip electric soldering iron for this, not a soldering gun.

When all LEDs are installed, replace the fuses in their respective spots. Be sure to observe the ampere ratings. Each LED should go off as you install its corresponding fuse. You can now drive in confidence, knowing that the LEDs are always on guard, ready to let you know before corrosion buildup causes trouble.

The LED idea is not limited to indicating fuse corrosion on VWs. It can be used on any car to tell you at a glance when a fuse or light bulb is burned out.

Fusible link is used to protect circuit that carries battery current to car's electrical system through ignition switch or fuse panel. It's easily identified by plastic tag that reads "fusible link." On factory cars, they're often wired between battery terminal of starter solenoid and main bulkhead connector, as shown here. In Ford solenoid shown on page 33, I used a 14-gage fusible link to protect a 10-gage wire feeding the electrical system, which is correct, as explained in text below. Photo by Jim Barrett.

Fusible links are sold in packages like this, complete with necessary fittings. Rating is printed on package. Fusible links are sold in bulk at some larger stores, but you'll have to provide fittings for them.

you're servicing or modifying an OEM harness, you may encounter one of these mysterious devices. I call them *mysterious* because few people know they exist, and fewer understand them.

Like any other fuse, the fusible link is designed for circuit protection. But in this case, the circuit being protected is the car's entire electrical system, or most of it. The fusible link is usually installed in the circuit between the battery and ignition switch, or between the battery and fuse panel, or both. Consequently, all the electricity used by the car, except to operate the starter motor, must flow through the fusible link(s).

A fusible link is simply a length of wire sized two gages smaller than the wire it's attached to, so it becomes the weakest link in the circuit. Remember, wire gages are only *even numbers,* so the fusible link is *four digits* smaller. This means that a 12-gage wire would be protected with a 16-gage fusible link. Aftermarket fusible links are often packaged with the notice: "This link protects a 12-gage circuit," or whatever is appropriate.

What distinguishes the fusible link from ordinary electrical wire is that its insulation is burn-resistant.

If a fusible link is overloaded, the conductor material melts, just as a fuse would. But, because the fusible link is usually hidden somewhere inside the wiring harness, its insulation must provide heat protection for the surrounding

wires. You can recognize a failed fusible link by its appearance. Although it won't melt, the insulation looks different—charred, blistered or discolored—in the area where the link melted.

To replace a fusible link, find the crimp attaching it to its wire. Often, one end is a ring terminal and the other is crimped to the wire. Seldom do you find a link spliced into the middle of another wire. Cut the crimp as needed and install a replacement link—each application has its own part number. Then make sure you find and fix the cause of the failure before powering up the system. Otherwise, you may be making a career of replacing fusible links!

Does your project car's electrical system need a fusible link? Hard to say. If you wire your car following the advice in this book, you will probably use more fuses in your car's electrical system than you'd find in a factory-built car, so overloaded circuits shouldn't be a problem. If this is the case, the only advantage of having a fusible link would be to protect the main feed wire to the ignition switch and fuse panel against a massive overload, such as a short-to-ground in the feed wire. However, fusible links are so readily available and so inexpensive that I think they are worth the extra effort to install. You can buy lengths of fusible link at any good auto-supply store, often with a ring terminal and crimp connector already attached.

CIRCUIT PROTECTION TIPS

Do not skimp on circuit protection. In other words, don't try to get by with only a few fuses or circuit breakers. Sure, you can use a high-rated fuse to operate three or four circuits all at once. But it's far better to give each major circuit its own correctly rated fuse or breaker.

Likewise, never under any circumstances design a circuit *without any protection.* If a short or overload occurs, the weakest link is going to burn, even if it's in the harness you so lovingly assembled.

Design your circuit-protection system so it can be easily added to later. The way to do this is to provide a fuse block with several extra sockets. This is far better than the makeshift method of using add-on fuse taps, which are good for little other than overloading existing circuits.

If you have repeated problems with fuses blowing in the circuit, don't arbitrarily install a higher-rated fuse—and don't bypass the fuse entirely, either! Remember, fuses don't blow without a reason. And if they blow repeatedly, the reason is serious enough to warrant careful diagnosis. The same holds true with circuit breakers.

On the other hand, a one-time blown fuse can be the result of something obscure, like a defective or marginally rated fuse, or some sort of transient condition in the circuit. If replacing the fuse cures the problem and it doesn't recur, don't get upset looking for a problem you'll probably never find.

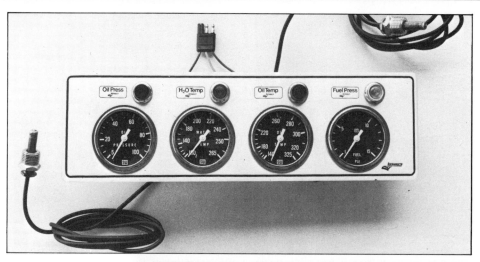

Some specialty suppliers offer prewired gauges mounted on flat inserts for easy mounting to different panel situations. This race-car/street-rod gauge setup, available from Longacre Automotive Racing Products, uses prewired Stewart Warner gauges mounted in labeled, brushed-aluminum panel. Photos by Jim Barrett.

Practically speaking, the instrument panel displays your car's vital signs every time you glance at a gauge or indicator lamp. How you use your vehicle largely determines the amount of information you need and in what form you need it.

For an embarrassingly long time, domestic auto manufacturers were convinced that none of their customers wanted to know any more than "how fast" and "how much gas." Everything else was monitored by indicator lamps, aptly dubbed *idiot lights*. Driving enthusiasts had only two choices—buy an import or add gauges to their American car. Somewhere along the way, Detroit got the message. Nowadays, you can find full instrumentation as standard or optional equipment on nearly every car line.

If you're adding gauges or other displays to an existing car, or building a car from scratch, the aftermarket offers a wide variety of choices, in style as well as in price and reliability. There's a list of major suppliers on pages 152-153.

This chapter describes the various systems and components you'll be dealing with when outfitting your car. It also includes enough theory and related information to help you make well-informed choices.

CHARACTERISTICS OF DISPLAYS

Anything that conveys information can be categorized as a display. However, in a car's electrical system, the three types most often used are *warning and indicator lamps, gauges and buzzers*. Displays depend on causing a stimulus, either to the driver's sense of sight (lamps and gauges) or sense of hearing (buzzers). The effectiveness of the device depends on its ability to intrude into the driver's consciousness and make an impression. Then, if he understands the

meaning of the impression, the driver can take the necessary action.

Regardless of the nature of the display device, you must consider several physical characteristics of the display when selecting one. As used in the following text, the term *discrimination* simply means noticing something and deciding how important it is. In the case of displays, there are several kinds of discrimination you need to consider when designing the instrument panel of your project car.

Visual discrimination of a display depends on its shape, size, color or indicator-needle position.

A good example of indicator position is the hands on a clock. You soon learn to tell time by the position of the hands, even if there are no numbers on the clock face. The same is true in a car's instrumentation. For instance, on an oil-pressure gauge you associate needle position with the part of the dial that's regarded as safe.

Color discrimination is often associated with warning and indicator lamps. It is also associated with the needle position on some gauges. That is, some gauges have color bands printed on the scale, so a quick glance shows you if the needle is in the red, green, or whatever. This way, you can tell status without looking at the numbers on the scale.

Red traditionally represents danger, so warning lamps, such as those used to indicate high temperature or low oil pressure, are often red. Conversely, indicator lamps that merely convey information are usually some color other than red. For instance, blue is frequently used for headlamp high beams, green for turn signals.

Auditory discrimination, of course, relates to hearing. In a car, a buzzer is either buzzing or it isn't. But you can discriminate between differences in the buzzer's *frequency* (high or low pitch) and its *intensity* (loud or soft), although these sound qualities aren't often important in a car's information systems.

The benefits of *auditory versus visual displays* are portrayed in the chart below:

Relative discrimination means comparing two or more stimuli to each other.

In a car, there's not much need for this, unless you happen to have temperature gauges for both engine coolant and engine oil. Here, a wise driver will make frequent comparisons, because a sudden rise in coolant temperature with little or no immediate change in oil temperature is a good early warning of coolant loss. For more information on interpreting gauge readings, see pages 46-48.

Absolute discrimination means taking a reading without comparing it to another reading. This kind of discrimination stands on its own merits. A simplified example is the GEN light used to indicate generator output on a car without a voltmeter or ammeter. If the light is off, you can presume everything is OK—if it's on, something's wrong. It's either one or the other, so there's no comparison needed.

Types of Displays—The information given by a display can be used to further describe it.

Static displays are those that never change. In a car, the only static displays in the electrical system are the identifying labels you may choose to place next to various controls, lamps or gauges.

Dynamic displays, on the other hand, are the ones that *do* change constantly. They are used to depict conditions of a variable. For instance, the fuel-level gauge constantly—sometimes disgustingly quickly—moves from FULL to EMPTY as you drive, to display the relative condition of the fuel level. Similarly, the voltmeter needle is constantly moving, usually in small increments, as the car's electrical system copes with changing loads. Also, a tachometer or speedometer changes with engine or vehicle

speed, respectively.

Although warning and indicator lamps don't actually move, you can think of them as dynamic, because they do have two modes—ON and OFF.

Quantitative displays are dynamic displays that present the *value* of a variable. For example, the fuel-level gauge just mentioned tells the *quantity* of fuel in the tank, either in gallons or in degrees of fullness.

Status displays present the condition of a system. The high-beam indicator and turn-signal indicator are displays of this type. So is the rear-window defogger ON lamp, and others of the simple ON/OFF type.

Warning displays are used to alert the driver of a dangerous or unsatisfactory condition. The GEN lamp, low-oil-pressure lamp and high-temperature lamp fall in this category. So do buzzers.

Representational displays are found on some newer factory-built cars, but are still uncommon in project cars. Honda automobiles probably are to be credited with introduction of the little pictogram showing the top view of the car with tiny color bars to indicate doors or hatch ajar. In my estimation, these displays have little more than novelty value. The same can be said of the lighted-bar graphs, currently in vogue in some instrument panels, to represent engine rpm, road speed and other vital information usually portrayed more clearly by conventional gauges.

INDICATOR LAMPS VS. GAUGES

First, take time to consider exactly what information you need the instrument panel to provide, and how you want it to be presented. Otherwise, you may end up with an instrument panel full of expensive overkill. "How come?" you might ask. "Aren't gauges regarded as an automotive *good thing?*" The answer is yes and no.

How much does your driving style demand that you know about your car's vital signs? How much do you already know, or are willing to learn, about interpreting gauge readings?

If you're still not sure of the answers to these questions after reading this section, you may be better off sticking with indicator lamps and spending the extra

DISPLAY CHARACTERISTICS

Auditory vs. Visual Displays

Characteristic	Auditory	Visual
Message	Simple	Complex
Duration of Message	Short	Long
Use of Information	Not referred to later	Referred to later
Driver Response	Immediate action	No immediate action
Preferred Location	Location too bright for visual display or with many visual stimuli	Location too noisy for auditory display or with many auditory stimuli

Race drivers usually dispense with speedo in favor of large, accurate tach with rev limiter. Shown here is the *RPM Limiting Pro Tach* from Mallory Ignition. It also has shift-point indicator lamp, set by dial at bottom right. Photo by Jim Barrett.

Race drivers also depend on large indicator lamps that quickly grab attention. These *Pro-Lites* from Auto Meter indicate low oil and fuel pressure; here, they're mounted on tach for best visibility. Photo courtesy of Auto Meter.

money on other accessories.

On the other hand, if you know enough about what makes a car tick to correctly interpret gauge readings, you probably do want to know what your oil pressure is at idle. You'll notice right away if a reading isn't quite normal. You'll watch to see if it was just a temporary fluctuation. And you know trends are important. Gradually dropping oil pressure may indicate a need to add oil, a failing pump, or some other condition that could get much worse before an indicator lamp would finally come on.

The needs of a race driver are different, too. For instance, lap times are important in track racing, while speed in miles-per-hour isn't. So, race cars usually don't have speedometers, but they do have a large, accurate tachometer positioned so maximum rpm is straight up. The tach may also have a *rev limiter* that prevents the driver from over-revving the engine.

The race driver is also concerned about oil pressure and temperature, coolant temperature and fuel pressure. If he's racing a VW-based vehicle, he'll want an oversized charge indicator because the cooling fan is driven by the generator belt.

So, what's best for you? Only you can answer that. For me, the optimum system for a street car contains: tachometer, speedo-land odometer-assembly with trip odometer, coolant-temperature

gauge, oil-pressure gauge, voltmeter, ammeter, fuel gauge, vacuum gauge and a clock—nine gauges in all. I'd also want indicator lamps for low oil pressure, charging system, high beam and turn signals. I've never seen a factory-built car with exactly this combination, so when I designed the instrument panel for my Bernardi component car, you can be sure my wishes were all met!

Indicator-lamp systems will be covered first, followed by a discussion of the various instrument systems, explaining what each gauge is for and how it operates.

INDICATOR LAMPS & SWITCHES

Regardless of which operating system is monitored by an indicator lamp, the lamp can tell you only two things—*yes* or *no*. Yes, the coolant is too hot. No, the oil pressure is not too low. No, the generator is not supplying enough current. The coolant-temperature and oil-pressure lamps are operated by on/off switches.

Indicator-Lamp Switches—Indicator lamps convey their information by lighting when something is wrong in the system being monitored. Because this requires nothing more than simple current flow, an indicator lamp can be controlled by an SPST switch. Indicator switches are usually called by the name of the system they monitor: oil-pressure switch, coolant-temperature switch, and so on.

When you look at an indicator-lamp switch, you'll see that it has only one terminal. This is because the switch is wired in series with the lamp and forms the ground end of the circuit. The circuit is completed to ground through the body of the switch. Battery voltage is available at the lamp; current flows only when the switch closes to provide a complete path to ground. You'll appreciate the logic of this when you wire your car and realize it would take twice as much wire if current went first to the switch, then to the lamp, and finally to ground.

Temperature switches rely on a temperature-sensitive *bimetallic element* operating a set of contact points. The switch is an SPST type, normally open. With increased temperature, the bimetallic element changes shape, causing the points to close at the *calibration temperature*.

The element is contained in the switch body and extends into the coolant to take a direct reading. By carefully choosing the shape, composition and tension of the bimetallic element, the designer can *calibrate* the switch, or determine its critical switch-point temperature.

The temperature switch resembles a short bolt, in that it has a threaded portion and a hexagonal head. The sensing element is housed in an extension beyond the threaded part. At the head, you'll find the single terminal that attaches to the ground side of the lamp.

American manufacturers use a standard-taper pipe thread (NPTF) for temperature switches. The taper ensures a pressure-tight seal without need for sealing compound, which would hamper current flow to ground.

Some European auto manufacturers also use the NPTF system, although the metric system is most common on popular imports. This isn't a problem with warning-lamp switches, though, because you'll likely stick with the factory switch anyway. Some special considerations for gauge sending units are discussed later.

Pressure switches are used for oil-pressure indicator lamps. In this application, the switch is a normally closed type—unless sufficient pressure is applied, the switch remains closed (on), providing a ground path for the warning lamp. In other words, oil pressure keeps

When I laid out instrument panel of my Bernardi component car, I made room for all instrumentation I wanted: nine gauges, five warning lamps and assorted switches. On the road, everything performs just the way I'd hoped.

Chevy V6 powered hot rod was scratch-built by Ron Fournier (author of HP's Metal Fabrication Handbook). Instrumentation is laid out just like on a race car. Although gauges are spread completely across dashboard, cockpit is narrow enough that even farthest gauge is within easy view.

the switch open (off) until the pressure drops below a critical level, closing it.

The pressure switch threads into a hole intersecting an oil galley in the engine. A small hole in the switch body permits pressurized oil to enter the switch, where it acts on a flexible *diaphragm*. Diaphragm movement operates a set of contact points that open or close the warning-lamp circuit, depending on diaphragm position.

The oil-pressure switch is calibrated to open the contact points above a specific oil pressure. Because oil pressure is largely proportional to engine speed, an idling engine will have normally low pressure, except when used in race cars. The switch is calibrated not to open at idle pressure, which would cause undue concern for the driver. But this characteristic is the major shortcoming of the oil-pressure-lamp system. The difference between *normally low idle pressure* and *dangerously low operating pressure* is narrow.

As a result of this narrow margin, many engines have been ruined because a warning lamp didn't come on until too late. I speak from a painful first-hand experience from my teenage years. An air-cooled VW engine sump holds less than 3 quarts of oil to begin with. My VW engine developed a gasket leak that drained all the oil in less than 200 miles of driving. The oil-pressure lamp came

on just in time to announce a completely fried engine!

Charge-Indicator Lamps—Of the three primary indicator lamps, this is the only one that operates without a separate switch. When you first start your car, the generator needs an external supply of current to get itself started—this is called *excitation*. Current, provided by the battery through the ignition switch, flows through the charge-indicator lamp on its way to excite the field windings in the generator. Naturally, with current flowing, the lamp lights. As soon as the generator begins generating current on its own, the voltage regulator allows some of this current back to self-feed the generator to keep it excited. When this happens, the indicator lamp goes out because it now has B+ on both sides, and current can no longer flow through it.

There's usually a resistor of about 15 ohms in parallel with the lamp. It carries a large percentage of the current necessary to excite the generator, allowing the lamp to carry a much smaller load. So, if the lamp filament burns out, the 15-ohm resistor is still in the circuit to carry current to the generator.

Guidelines for Selecting Indicator & Warning Lamps—Size, brightness, and the period of time available for observation are all important considerations when choosing indicator or warning lamps. Lamp colors that most

Location of oil-pressure warning lamp switch varies from engine to engine, but is always in an oil galley. On this 2.3 liter Ford engine, it's at the rear, high on the head. Switch takes reading as far from oil pump as possible, to indicate that there's adequate pressure throughout the system.

quickly attract attention are, in order: red, green, yellow and white.

In addition to lamp color, background color and the amount of ambient light also affect visibility. Obviously, even a red lamp is not going to stand out well against a red instrument panel on a bright, sunny day. So consider these points: For best visibility, choose a lamp color that contrasts with its background. If the lamp has good contrast with the background, and ambient light is very bright, the warning-lamp *color* is less important than its relative *brightness*. When contrast between lamp and back-

Charge-indicator lamp (left) illuminates dimly as the charge rate drops toward zero; glows brightly when generator output is zero. My car also has voltmeter and ammeter, so it's not serious that indicator lamp is partially obscured by steering wheel rim.

Examples of analog gauges. Although oil pressure gauge provides a psi scale that gives useful information, it is "hidden" at bottom of tach, rather than being in a more prominent spot. Ditto voltmeter.

Mechanical gauges—especially fuel gauges—should be mounted *outside* driver's compartment. Photo courtesy Auto Meter.

ground is not strong, follow the color order just listed for best detectability.

GAUGES & SENDERS

Unlike an indicator lamp with its yes/no message, a gauge tells you *how much.* Generally, the information is displayed by a *movable needle* pointing to a calibrated scale. This type is often called an *analog gauge* because the needle position and numbers on the gauge correspond to directly measurable quantities. That is, so many degrees of needle movement represent so many pounds per square inch, revolutions per minute, degrees of temperature, and so on.

A second type of gauge has recently become fashionable. It's called a *digital-readout* gauge because it presents information directly in numbers. Digital gauges are used almost exclusively in high-priced factory cars, although some are becoming available in the aftermarket.

You may find digital gauges more difficult to read than the analog type. With an analog gauge, you soon become so familiar with the normal needle position that you can read the gauge without actually looking at the numbers. For proof of this, think about the designer clocks you've seen that don't have numbers on the face, yet you can tell time on them, even at a glance, by the position of the hands. With the analog gauge, you can frequently and instantaneously check your car's condition.

This is not so easy with digital readouts. Because they can present only one number at a time, you must consciously read and *interpret* the numbers each time you consult them. These gauges don't show an overall scale that you can readily compare the isolated readings against. This can divert your eyes and attention from the road for an uncomfortably long time. Besides, what good does it really do to know that your coolant temperature is *exactly 181F?* Or that your oil pressure is *exactly 58 psi?* Also, studies in the field of *ergonomics* have proven that analog gauges are easier to read than digital gauges. If you get the idea that I really don't like digital readouts for normal use, you're right!

On the other hand, they're great for diagnostic equipment, such as monitor-ing and recording test values. An example would be thermocouple-gauge readouts when you're monitoring exhaust-gas temperatures.

Types of Gauges—Both OEM and aftermarket gauges fall into one of two major categories—*mechanical* or *electrical.* Because this is a book about electrical systems, mechanical gauges are only briefly described here. Most gauge manufacturers offer both types—you can contact them for additional information on their mechanical gauges.

Mechanical gauges for temperature and pressure require plumbing in the form of tubing to transmit information directly to the gauge. For instance, mechanical temperature gauges utilize a sealed bulb and tube (capillary tube) somewhat like a medical thermometer. The gauge needle is moved by the expanding fluid inside the tube. The tube must not be sharply bent and must be guarded against breakage. In my opinion, these requirements seriously limit installation convenience.

Pressure-operated mechanical gauges require that the pressurized liquid actually be piped into the back of the gauge. Leakage is always a concern. In fact, *mechanical fuel-pressure gauges must never be mounted inside the passenger compartment* because of the potential fire hazard. Because these gauges are usually mounted in the engine compartment, their practical use is usually limited to out-of-car readings. They're sometimes used on race cars, mounted outside the passenger compartment within the driver's view. Mechanical oil-pressure gauges are acceptable, though uncommon, in passenger-car instrument panels.

Mechanical gauges do have one distinct advantage over comparable electric gauges. The mechanism of the gauge is not limited to the 90° needle movement of electrical gauges. In fact, a 270° swing is common on mechanical gauges. Greater needle movement translates directly to more accurate interpretation of readings. Also, the needle must move farther to get from the safe zone to the unsafe zone, which means that your eye will more readily catch an incorrectly positioned needle. This is the main reason race-car drivers prefer mechanical gauges.

Some gauges are available in mechanical form only. These include vacuum gauges and turbo-boost gauges. Most speedometers are mechanical, although some of them have been electrified successfully. You need to consider this if you're building an instrument panel from scratch. Some manufacturers have a relatively limited selection of speedometers and other mechanical gauges, so matching all your gauges for style may take some extra effort.

Electrical gauges and their sending units are designed to convert a physical condition into an electrical signal that can be transmitted over an ordinary electrical wire, then convert the electrical signal into a meaningful visual display.

A big advantage often cited in favor of electric gauges is that corresponding mechanical gauges require hydraulic plumbing, capillary tubes, mechanical drive cables, or other difficult-to-route and space-consuming transmission devices.

Electrical gauges operate by several principles, but the two most popular by far are the *thermostatic principle* and the *electromagnetic principle,* also called the *balancing coil* or *air-core* principle.

Thermostatic gauges are found only in factory installations. I have not discovered examples of this type in the aftermarket except as OEM replacements. The balancing-coil gauge is the only kind sold for aftermarket installations and is used extensively in factory installations, as well.

Thermostatic Gauges—Thermostatic gauges work on a bimetal principle. A strip of two dissimilar metals bonded together is anchored at one end and is attached by linkage to a movable pointer at the other. Because the two metals expand and contract at different rates, the bimetallic strip will bend when heat is applied, causing a corresponding movement of the pointer.

The heat needed to operate the pointer is provided by a small heating coil wrapped around the bimetallic strip. Current flow through the heating coil is regulated by a variable resistance in the sending unit, as discussed later. So, as current flow varies, the heating coil changes temperature, bending the bimetallic strip, which moves the pointer.

By calibrating the variables, the gauge designer can make the pointer position correspond directly to whatever physical condition is causing a resistance change in the sender. One side effect of the direct link between the bimetallic strip and the pointer is that when the engine is shut off, the pointer swings to the low position as soon as the bimetallic strip cools, and stays there.

Because heating-coil output in a thermostatic gauge is related directly to voltage, a potential problem could develop. If the gauge operated on 12 volts, and system voltage were low, it would give an incorrect reading. To counteract this, thermostatic gauges are designed to operate at a voltage lower than 12 volts— about 7 or 8 volts.

A small *constant voltage regulator* (CVR) reduces whatever the system voltage is at the moment to a constant gauge voltage. This ensures relative accuracy unless system voltage is dangerously low to begin with. In that case, you have problems to deal with other than worrying about gauge accuracy!

Cost and complications of the CVR and its requisite radio-noise-suppression filter are two reasons aftermarket gauge manufacturers and many auto manufacturers have gone entirely to the balancing-coil-gauge system.

Balancing-Coil Gauges—The main advantage of balancing-coil gauges is they operate accurately on whatever battery voltage happens to be, without need of a CVR.

Inside a balancing-coil gauge, there are two electromagnetic coils wound around a plastic core at right angles to each other. The pointer shaft runs between the two coils and has a metal armature on which both coils can act. When the system is shut off, there's nothing to act on the pointer, so it assumes a *random position* until the system is turned back on. This, of course, distinguishes the balancing-coil gauge from a thermostatic one.

The two coils are wired in series, so current flows first through one coil, then the other coil, and finally to ground. The two coils are connected to terminals on the gauge case, which are marked B, I or + for the feed side and G or − for the ground. Check the gauge instruction sheet for exact markings. A third wire from the sending unit taps into the circuit between the two coils and connects to a third terminal, usually marked S for sender.

The sending unit used with balancing-coil gauges must be capable of offering a variable resistance in direct proportion to changes in the system being monitored. Let's use the fuel-level gauge as an example of how a typical balancing-coil gauge works.

For this discussion, the gauge coil closest to the battery feed will be called the *E coil* (for empty), and the one closest

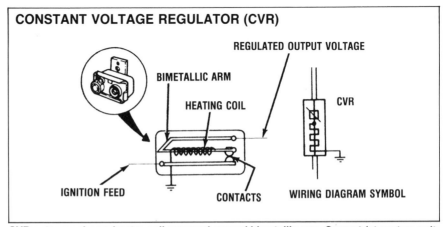

CONSTANT VOLTAGE REGULATOR (CVR)

REGULATED OUTPUT VOLTAGE

BIMETALLIC ARM

HEATING COIL

CVR

IGNITION FEED

CONTACTS

WIRING DIAGRAM SYMBOL

CVR cutaway shows heater coil wrapped around bimetallic arm. Current (at system voltage) flows through heater until arm bends to close contacts. System voltage is applied through points. After a moment, arm cools enough to open points, and cycle repeats. The short bursts of electricity through points are so close together that they average out at less than system voltage. Drawing courtesy of Chrysler Corp.

This CVR is supported by gauge cluster assembly and is attached to printed circuit by snaps. When I built my car, I discarded this whole assembly and installed VDO gauges instead.

Sender for oil-pressure gauge is physically larger than temperature sender because it contains a presssure-sensing diaphragm.

BALANCING-COIL GAUGE

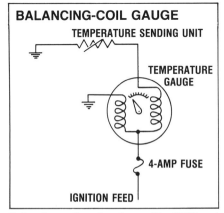

Balancing-coil system is used by OEM and aftermarket gauge manufacturers. Two magnetic coils receive equal voltage through fuse. One coil is constantly grounded; other is grounded through variable resistance of sender. Differences in magnetic field cause needle to move.

THERMOSTATIC GAUGE

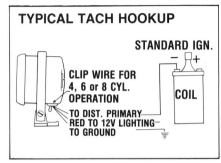

Thermostatic gauge has heater coil wrapped around bimetallic arm. Arm is linked to needle. Current flow through heater is controlled by variable resistance in sender. Varying current also varies temperature of heater coil. Bimetallic arm bends and moves needle accordingly.

TYPICAL TACH HOOKUP

Diagram shows typical tach installation, wired to battery negative side of coil. There are exceptions, so check car's service manual and tach manufacturer's instructions. Drawing courtesy of Auto Meter.

to ground the *F coil* (for full). When current is applied to the coils, the E coil creates a magnetic field that tries to attract the pointer armature to it. But the F coil also creates a field that counteracts the E field. Current flow through the E field remains constant, while the strength of the F field is governed by the variable resistance of the sender.

When the fuel tank is full and sender resistance is high, current tends to flow through the F coil, strengthening its field, so the pointer indicates FULL. When fuel level goes down, so does sender resistance. Now, current flows where resistance is lower—through the sender—and current flow through the F coil drops. This weakens the F field, and the pointer armature is attracted toward EMPTY. Pressure and temperature gauges work similarly.

Tachometers—One characteristic common to all electrical tachometers is, although they may not be attached directly to it, they're operated by a signal from the ignition coil. This signal is then processed electronically into needle movement representative of engine rpm.

There are several things to consider when selecting a tachometer. First, you must match the tachometer to the number of cylinders in the engine. Some tachs are designed for a specific number of cylinders. Multi-use tachs are designed with a selection of terminals that correspond to the number of cylinders, or a tiny dial on the back that serves the same purpose.

You just hook up the feed wire to the appropriate terminal or move the dial to the desired position.

Second, be sure the tach you buy is compatible with the voltage of your car's electrical system.

Third, electrical tachometers don't require a sending unit. As mentioned earlier, they are wired directly into the ignition circuit, usually on the battery negative (−) side of the coil. With the older point-type ignition systems, this is a straightforward connection. However, follow the wiring instructions that come with the tachometer.

On electronic-ignition systems, consult the OEM shop manual for factory ignitions or the instructions that come

with the aftermarket ignition. Also, discuss compatibility with the auto-parts dealer or the tach manufacturer. Installation directions vary too much to list all variables here.

Finally, some highly specialized aftermarket tachometers come with a built-in *rev limiter* that shuts off the ignition coil for a brief period if a pre-selected rpm is exceeded. This protection device is worthwhile for race drivers whose driving conditions may make them more prone to accidental over-revving. If your street driving includes this kind of antisocial behavior, the rev limiter may be just right for you, too! Special wiring considerations for rev-limiter tachometers are supplied by the manufacturer.

Sending Units—You've already seen how electrical gauges interpret the variable resistance in a sending unit to display the car's physical conditions. Operating principles of sending units are discussed here.

Electric gauges operate on varying resistance in the sender circuit. Sending units convert a physical condition—pressure, temperature or fluid level—into a specific electrical resistance that is interpreted by the gauge head. As with indicator lamps, battery voltage is applied first to the instrument, then flows through a single wire to ground through the sending unit.

Temperature sending units rely on an electrical element that varies its resistance in proportion to temperature. Most electrical conductors display an increase in resistance as temperature rises. However, there's one unique family of conductors that behaves just the opposite. They're called *thermistors,* and their resistance *decreases* as temperature rises. Thermistors are made from the oxides of cobalt, copper, iron or nickel. The advantage of using a thermistor as a sending unit is that it is a solid-state device—no moving parts to go out of order—and calibration is built in when the thermistor is compounded.

Here's how the thermistor functions as a sending unit: When temperature rises, thermistor resistance drops. This allows more current to flow through the thermistor and the gauge, moving the gauge needle upward on the temperature scale.

Other temperature sending units rely on a specialized rheostat inside the sender housing. A bimetallic strip or coil changes shape with variations in temperature. Mechanical linkage converts bimetal movement into rheostat-wiper movement, which provides variable resistance. Because they're more prone than thermistors to lose calibration or otherwise malfunction, the rheostat sending unit has fallen into disfavor. You're likely to see one only on an older car. They're easily distinguished from the thermistor type because of their large size.

A coolant-temperature sending unit usually threads into a hot-water passage in the engine. Most of these sending units come in a variety of standard thread

COOLANT-TEMP. SENDER

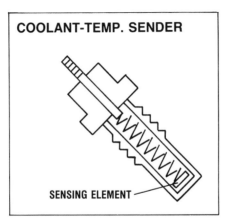

The thermistor-type coolant-temperature sender contains sensing element held in place by electrically conducting spring. Changes in temperature change conductivity of sensing element.

Here's how I installed oil-pressure gauge and warning lamp in my car. Brass T-fitting is threaded into oil galley to provide two openings—one for each sender. Because gauge sender is the heavier of the two, I suspended it vertically to minimize vibration.

FUEL-GAUGE SENDER

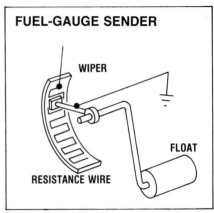

Note similarity between this fuel-gauge sender and oil-pressure sender below. Major difference is that wiper is moved through linkage attached to a float.

OIL-PRESSURE SENDER

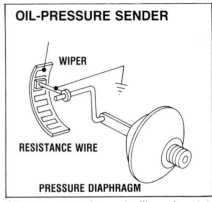

Oil-pressure sender works like a rheostat. Wiper, moved by pressure diaphragm, slides across coil of resistance wire, completing circuit. Position of the wiper on coil determines how much or how little resistance sender offers.

sizes. Metric adaptors are available for most.

Special threaded adaptors may also be needed to install temperature senders to monitor the temperature of transmission fluid or axle lubricant. For transmission fluid, install the sensor in the fluid-cooler return line, if possible. Refer to the car's shop manual to positively identify the return line.

In the case of oil-temperature sending units, you can install the sensor in a newly drilled hole in the block or oil pan below crankcase-oil level. Or, you can use a special sender as a replacement for the original oil-pan drain plug. In dry-sump systems, it should go in the bottom

of the oil-reservoir tank. In any case, be sure the location doesn't interfere with internal moving parts, obstruct normal oil flow or change road clearance.

Pressure sending units are physically larger in size than thermistor temperature units because they must enclose a diaphragm, linkage and a rheostat. Like the pressure-indicator lamp switch discussed earlier, the sending unit for the oil-pressure gauge threads into an appropriate hole in the oil galley.

In the sending unit, oil pressure acts on a flexible diaphragm, moving it in proportion to changes in pressure. A system of levers attaches the diaphragm to the movable wiper of a calibrated rheostat.

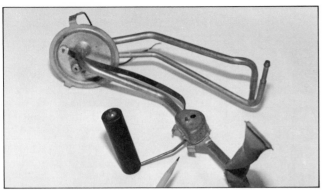

Pencil is pointing to variable resistance coil inside fuel-level sender. When unit is installed in fuel tank, black float will float on top of fuel. Movement of float varies resistance of sender.

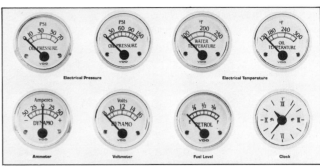

Most gauge manufacturers offer several full lines of gauges in several styles. VDO *Vintage* series might be good choice for antique or classic replicars. Photo courtesy of VDO Instruments Inc.

When oil pressure is low and the diaphragm is relatively relaxed, resistance is low. As pressure increases, so does resistance.

If space permits, it's possible to use a T-fitting in the engine, with both a pressure sending unit and oil-pressure lamp switch threaded into it. I used just such an installation on my car so I could have both an oil-pressure gauge and oil-pressure warning lamp. Various other thread adaptors are available, including metric sizes.

Fuel-level sending units go inside the fuel tank, often attached to the fuel pick-up tube. The sending unit is a type of rheostat. A float is attached to a movable arm. The movable arm moves a wiper along a coil of resistance wire. The float position, governed by fuel level, determines how much resistance wire the fuel-gauge current must pass through. Varying resistance in the circuit is interpreted by the fuel gauge.

Guidelines for Selecting Gauges—For best detectability, use a gauge with a moving needle and a fixed scale, called an *analog gauge*. This type capitalizes on the idea mentioned earlier about becoming familiar with needle position. Studies have shown that you can read this type of gauge faster than any other types, even though it may not look quite as "high-tech" as the digital or lighted-bar gauges.

When selecting a gauge, be sure the scale (numbers) covers only the range of values you expect to encounter. As an example, the car's voltmeter should read from 0 to about 16 volts. Don't bother with one that reads much higher, because

14.7 volts is about the maximum you should expect to see in a normally operating generating system. Choose an oil-pressure gauge that reads slightly higher than the maximum calibrated pressure of your car's oil-pump relief valve. Selecting a gauge with a higher-reading scale is simply wasting part of the gauge. More importantly, it condenses the area you're concerned with into a too-small space, so readings are less precise. The tachometer is one example where many people tend to go overboard. Let's face it, will your engine really reach *9,000 rpm* without turning it into junkyard material? You're far better off getting a tach that reads just slightly above your engine's red line. Anything higher simply wastes gauge space and doesn't impress too many people.

ANALYZE YOUR ENGINE AS YOU DRIVE

Stewart-Warner Instrument Division published an excellent brochure under this title several years ago. It tells how to use the various gauges in your car to analyze engine performance. They kindly granted me permission to share with you some of the ideas presented in it.

Efficiency is the key word when you talk about engine performance. Whether your interest is fuel economy, clean air or quarter-mile times, your best chance of achieving your goal is with an efficient engine. And maintaining that efficiency demands adherence to a set of operating standards.

An engine can be finely and correctly assembled and adjusted in a garage, but

on the road it's in a constant process of deteriorating out of specification. So, you need a way to monitor the ever-changing conditions, a way to discover and interpret trends, and a way to catch problems while they're still manageable. The gauges on your instrument panel offer this kind of information, if you know how to use them.

Stewart-Warner's first suggestion makes excellent sense: "If your car came with inadequate instrumentation, install the gauges while your car is running right." When the car is new or freshly rebuilt would be the best time to do this. This way, you'll be able to learn what readings to expect when conditions are normal for your particular car.

Their second suggestion is to write down the normal readings under a variety of conditions for reference later, as the car ages and loses its initial edge. They suggest taking separate readings with outside temperatures of 20F (−5C) and 70F (21C). Obviously, where you live will have some bearing on the temperatures you choose. The idea is to be able to recognize what readings are normal for the temperatures you will encounter while driving.

Record gauge readings with the engine idling, at city speed, at highway speed and while pulling a load or climbing a grade, or under race-car conditions. For ammeter and voltmeter, note readings with ignition switch OFF, ON (not running), and START (cranking). Also take readings with engine idling and at the road speeds mentioned above. Additionally, note the ammeter reading for each accessory used separately.

The following information will help you interpret your gauge readings. It also explains what to do with the interpretations.

Tachometer—The tach reads the engine revolutions per minute (rpm): Use it to set idle correctly. Depending on your vehicle, this helps avoid stalling, excessive wear on automatic-transmission clutches and excessive fuel consumption. Use the tach to judge when to shift for optimum economy or performance and to avoid lugging the engine.

The tachometer can also be used to check for clutch or transmission slippage. In a specific gear, road speed versus rpm should always be directly proportional with a manual transmission. A heavily loaded automatic transmission may exhibit normal torque-converter slippage. This is a sign to carefully monitor all temperature indicators for early signs of overheating.

Oil-Pressure Gauge—Indicates oil pressure in pounds per square inch (psi): There is no one oil pressure that's right for all engines and conditions. Engine-oil pumps are mechanically driven, usually by the camshaft or distributor, or, in a dry-sump system, a toothed belt off the crank. This means pump speed and resulting oil pressure are somewhat directly related to engine rpm, within the maximum limits of the pressure-relief valve. When an engine is new and everything is operating correctly, oil pressure will be very low at idle—say 20 psi—and will increase as engine speed goes up, to a regulated maximum—say 60 psi for a typical street engine. 80 psi is not uncommon on a race-car engine.

When the engine is cold, expect to see higher oil-pressure readings because the oil is thicker. As temperature rises, oil pressure should stabilize at the normal readings.

Low pressure while driving indicates a problem related to inadequate lubrication. Suspect low oil level, incorrect oil viscosity, overheated engine, clogged oil filter, oil foaming, or failure of engine bearings.

Oil-Temperature Gauge—Indicates oil temperature in degrees Fahrenheit (F) or Celsius (C): Oil temperature is an indication of engine load because oil comes into direct contact with the engine's moving parts. It also absorbs heat through the block and heads from engine coolant. Expect higher temperatures when your engine is at or near wide-open throttle for a sustained period.

Other causes of higher-than-normal oil temperature include low oil level, incorrect timing, incorrect spark-plug heat range, and cooling-system problems.

Coolant-Temperature Gauge—Indicates coolant temperature in degrees Fahrenheit (F) or Celsius (C). Coolant temperature and oil temperature don't indicate the same things because they're sensed differently. As mentioned, oil temperature is derived more from moving engine parts. Coolant temperature is related more to the effects of combustion. Also, coolant is separated from the direct source of combustion heat by the thickness of the coolant-passage walls.

Older engines were designed to operate in the 140—160F (60—72C) range, controlled by an appropriate thermostat. Modern production engines routinely run closer to 200F (94C), because of emission concerns. This would be dangerously close to the boiling point of water—212F (100C)—except for two things: Pressure radiator caps and antifreeze allow higher boiling points.

You can see in the accompanying chart that in the newer cars that use a 15-psi cap and recommended 50/50 coolant mixture, the coolant temperature has a comfortable safety margin.

Engines often exhibit a phenomenon called *afterboil*. This occurs when the engine is turned off and coolant circulation stops. In localized areas, noncirculating coolant may absorb so much heat from the surrounding passages that it temporarily rises above the normal maximum temperature. You may notice this indicated on the temperature gauge if you turn the ignition key back on after the engine stops. Afterboil is not considered abnormal as long as the radiator cap maintains system pressure and does not permit steam or coolant to escape.

If an engine regularly overheats or loses coolant, you may be able to use the coolant-temperature gauge to diagnose the cause. If, for instance, boiling occurs at a temperature lower than shown in the accompanying chart, the ethylene glycol concentration may be too low. Or, your engine may have a pressure leak in the coolant system or head gasket.

When pulling a trailer or other heavy load with an automatic-transmission-equipped vehicle, you can expect higher-than-normal coolant temperatures. This happens because the transmission-oil cooler, located in the radiator tank, transmits excessive heat into the coolant.

When you cold-start an engine, pay attention to the temperature-gauge reading. It should begin rising after a few minutes and should continue to rise until it reaches the normally indicated reading. If rises too quickly, you may have a low

RADIATOR BOILING POINTS
(Maximum Temperatures)

Open Cap	15-psi Cap	Coolant Type
212F	250F	Plain water
226F	268F	50% water, 50% glycol
240F	240F +	100% glycol

Tach reads revolutions per minute (rpm). Sun Super Tach II, model CP7901, has adjustable "red line" to indicate optimum shift points. Photo courtesy of Pentron Products Inc.

coolant level, failed water pump or blown head gasket.

Cylinder-Head Temperature Gauge —Indicates cylinder-head temperature in degrees Fahrenheit (F) or Celsius (C): Because air-cooled engines rely on air instead of water for cooling, it's not practical to measure operating temperature directly from the coolant (air). Instead, a special gauge is used to indicate the temperature of the cylinder head. Temperature ranges encountered by the gauge sensor are much different than those in a water-cooled engine. Ranges between 100F and 600F (40C and 320C) are not uncommon, with 250—400F (120—200C) regarded as normal.

A temperature above the acceptable maximum may indicate low oil level, cooling-system problems, incorrect timing or an excessive load on the engine.

Ammeter—Indicates the flow rate of current into or out of the battery: The ammeter is calibrated in amperes for both charge (+) and discharge (−) rates, in ranges of 20, 30 or more amperes. The larger the range, the less precisely you'll be able to read the ammeter scale.

In an absolutely ideal electrical system, the ammeter would read 0, indicating that generator output is exactly equal to electrical demands at the moment. A high (+) reading indicates that, at the moment, the generator is providing extra current to recharge the battery. This is normal immediately after engine start-up.

If an overcharge condition persists, there are two possible causes. The bat-

This cylinder-head temperature gauge indicates head temperature in degrees Fahrenheit. Others indicate temperature in Celsius. Photo courtesy of VDO Instruments Inc.

tery may no longer be capable of taking a full charge, even though the generator is trying to provide it. Or, the voltage regulator may be out of calibration. Testing will be needed to pinpoint the problem. A moderate, continuous charge (+) reading indicates a trickle charge from the generator to maintain battery condition. This is normal and is not cause for alarm.

A low (−) reading indicates that, at the moment, electrical demands are more than can be met by the generator, and the battery is discharging to make up the difference. DC generators used in older cars will display this condition at idle, which is not desirable, but normal. Incidentally, during engine-starter operation, current used to run the starter motor does not flow through the ammeter and, consequently, isn't registered by it. Current used to operate the starter relay and ignition system is indicated.

If a low (−) reading persists at normal engine speeds, suspect inadequate generator output. Possible causes include a loose drive belt, incorrect calibration of the voltage regulator, an undersize generator, or failed components in the generator or regulator.

Other causes of a constant discharge could be a short circuit or an abnormally high current draw in one or more accessories. Selectively switching accessories on and off may help locate the fault.

Voltmeter—Indicates system voltage: When the engine is not running, the voltmeter indicates battery voltage, which should be 12—13 volts in a 12-volt system or 6—6.5 volts for a 6-volt system. When the engine is running, the voltmeter indicates total system voltage, which is the combined voltage of the generator and battery. Generally, total system voltage should be 13.3—15.2 volts.

A high voltage reading may indicate incorrect calibration of the voltage regulator. If you don't correct this condition, you can expect shortened battery life, caused by overheating and possibly boiling. Excessive voltage also shortens the life of lamp filaments.

A low voltage reading may indicate incorrect regulator calibration, loose belts, a defective generator or excessive electrical demands. A temporary low

reading—as low as 10 volts—is normal during engine cranking, especially during cold weather.

Because the ammeter and the voltmeter give instantaneous information, you'll need to interpret a number of individual readings over a span of time to determine meaningful trends.

Neither meter is "superior" to the other, although there are some who will argue in favor of one or the other. If pressed for a choice, I'll admit favoring the voltmeter, for the simple reason that a generator warning lamp tells me all I need to know about charge/discharge. The voltmeter gives more meaningful information about *battery* condition. I installed both meters in my car.

Fuel-Level Gauge—Indicates relative fuel level in the fuel tank: Most fuel-level gauges are calibrated, not in gallons, but in relative amounts of tank fullness. When you become familiar with your car, you'll be able to assign approximate gallon amounts to needle position.

Some fuel-level gauges are equipped with special features that deserve mention. On vehicles with *multiple fuel tanks,* you may find a separate gauge for each tank, or a selector switch that operates a fuel-flow solenoid valve and fuel gauge for the tank selected.

Some OEM gauges include a *low-fuel-level warning lamp* that operates electronically from the conventional fuel-level float in the tank.

Electric Speedometer—Indicates road speed in miles per hour and distance traveled in miles: Most speedometers are mechanically operated by a drive cable, so the electric type is somewhat of an oddity. A sensor driven by the transmission sends impulses corresponding to road speed. On early VWs, this sensor is driven by one of the front wheels. The electric-speedometer head translates the impulses into road speed, which is indicated by the needle. Simultaneously, the *odometer* translates the impulses into distance traveled. One electric speedometer, offered by Classic Instruments, features an odometer that can be custom-calibrated to the gear ratios and tire size of any specific car. This is an attractive feature to those who constantly experiment with gear ratios and tires in search of better performance.

Electrical Loads

Lights are resistance loads. Headlamps and driving lamps on this Class 7 4X4 off-road racing truck are *heavy loads* that draw high current, so they deserve special consideration when selecting wire, circuit protection, battery and alternator. Photo by Tom Monroe.

Electrical loads are the devices that perform beneficial work in a car's electrical system. They're called *loads* because they act as burdens to the battery and generator, drawing current as they do their work. As mentioned in chapter 2, loads rely on one of two principles to do their work: *resistance* or *magnetism*. The application of these principles is explained in the following text. This is followed by a detailed explanation of the three basic kinds of loads found in an automotive electrical system—lamps, ignition systems and motors.

RESISTANCE LOADS

In an electrical circuit with no resistance, amperage would be incredibly high. The excessive current flow would create heat, which would quickly melt the conductor. On the other hand, if resistance is too high, little or no current can flow. Resistance loads strike a medium between these extremes, harnessing heat to do useful work.

Resistance loads are designed to do one of two things: generate *heat* or create *light*. In automotive applications, two common *heat-generating loads* are the rear-window defroster and cigarette lighter. Some old cars, like the 1953 MG-TD I once owned, had a heating strip that attached to the windshield with suction cups, to defrost the glass. It didn't work well!

In a heat-generating load, the ohm value of the resistance is carefully selected so current flowing through it creates the desired amount of heat. In other words, current flow actually creates the heat, but resistance controls the amount.

The resistance material, often called the *element,* must be thinner than the electrical wire serving it so the element gets hot but the wire doesn't. In this respect, the element is similar to the element in a fuse, with one notable exception. The heating element is made of a material that allows it to become hot

without melting. Alloys of *tungsten* or *nichrome* are commonly used for heating elements.

Because heat-generating loads draw high currents, the fuses and wiring used with them must be carefully selected. Also, keep in mind that resistance loads can quickly drain a battery if generator output is inadequate.

Light-generating loads (lamps) are the most common electrical loads in a vehicle. There may be a dozen or more on the outside and two to three dozen more inside. Like heat-generating resistance loads, lamps have an element, called a *filament,* that is much thinner than the wire serving it. When current flows through the filament, it gets so hot that it produces visible light. Because lamps are so numerous and so important in the electrical system, they are discussed in detail here.

LAMPS

Lamps consist of three basic com-

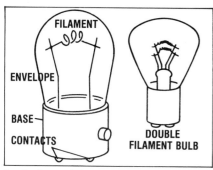

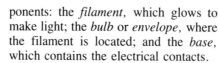

In single- and double-filament light bulbs, filaments are suspended near center of glass envelope for maximum light dispersion. Also note how contacts protrude from base in these automotive bulbs.

Left and center lamps have bayonet bases. Right lamp is installed simply by plugging into its socket. Contacts are lengths of wire that protrude directly through glass.

ponents: the *filament*, which glows to make light; the *bulb* or *envelope*, where the filament is located; and the *base*, which contains the electrical contacts.

Lamp configurations are as varied as the specialized uses they are designed for. Some, like headlamps, have a built-in *reflector* to concentrate the light beam; others don't. Some have a metal base; others rely on the shape of the glass bulb for mounting. Some lamps, such as headlamps and taillamps, have two filaments, while others have only one. On some lamps, the filament protrudes through the bulb to form connector terminals. Other lamps either use the metal base as a terminal or have spade terminals embedded into the glass bulb.

As mentioned, the filament is the part that actually produces the light. It works on the same current-flow/resistance principle as the heat-generating loads discussed earlier. The filament is sealed inside an oxygen-free atmosphere. This permits a great deal of current to flow through the filament relative to its size—wire diameter—so it becomes hot enough to produce visible light.

The filament usually consists of an extremely fine strand of tungsten wire wrapped into a tight coil to provide a greater amount of light-emitting surface. The coil is supported by heavier wires inside the bulb, where it's relatively safe from damage by impact.

Tungsten has several characteristics that make it desirable as a filament—it gives off intense light when heated, and it's fairly rugged. But, it also has drawbacks. In a conventional automotive lamp, the light it produces is not really white. Although an untrained eye may not see it, tungsten light is yellow. Also, tungsten tends to slowly evaporate when heated. The resulting tungsten vapor forms a film on the inside surface of the bulb, gradually reducing light output. Eventually, the filament becomes thin enough to burn out or break from road shock. Then the bulb must be replaced.

The bulb is made of heat-resistant glass. During the manufacturing process, air is pumped out of the bulb and is replaced by an inert gas, commonly *argon*. Without oxygen, the filament lasts for an extremely long time without burning out. In the presence of oxygen, the filament lasts only a fraction of a second because it burns up. If you've ever accidentally broken a lamp bulb while it is on, you've seen this happen.

The *base*, or *mount*, of a lamp is an important consideration when designing your electrical system. It isn't that one kind of base is necessarily better than another, but because each type of base can accept a specific kind of lamp socket.

Depending on the number of beams, a headlamp has two or three *spade terminals* extending directly through the

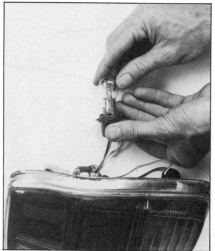

Bayonet base of single-filament taillamp bulb must be inserted into its socket with locating pins in correct channels, then twisted slightly to lock in place. Photo by Jim Barrett.

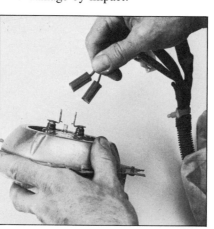

Single-filament headlamp has two spade terminals—one to conduct current, one for ground. Terminals for both single- and double-filament headlamps are standardized for universal fit. Lamp can use individual female connectors as shown here, or single molded connector with two sockets. Photo by Jim Barrett.

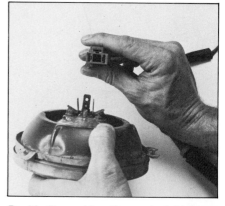

Double-filament headlamp—both high and low beam in one lamp—uses three terminals, one for each filament and one for ground. Factory cars use one-piece "foolproof" connector, like one shown here. In a hobby car, you can also use individual female connectors like those shown for single-filament lamp at left. Photo by Jim Barrett.

glass bulb. Low- and high-beam lamps each require a specific connector. One terminal is for ground. The others conduct current to the filament or filaments used for high or low beam.

Turn-signal lamps and taillamps usually have a metal *bayonet* base with small pins used to attach the base to the socket. On single-filament bulbs, the pins are equidistant from the end of the base. On multi-filament bulbs, the pins are located at two different distances so they can be inserted into the socket only one way. If a bayonet lamp won't easily twist into place, remove it and rotate it 180° to align the pins the opposite way. Don't force it.

In OEM applications, bulbs used for instrument-panel illumination have no separate base. Instead, the glass is shaped to snap into the grippers of the socket. Filament wires extend through the glass and are trapped between the glass and the socket grippers (contacts). Don't twist this kind of bulb because the metal grippers are delicate.

Halogen Lamps—The ideal headlamp produces brilliant white light. It doesn't lose intensity because of tungsten deposits. And, it rarely burns out because of a weakened tungsten filament. The conventional sealed-beam headlamp falls short in all these areas. Enter the *halogen lamp*.

The halogen lamp is also referred to as a *quartz-iodine* lamp, or *quartz-halogen* lamp. In any case, the quartz part of the name refers to the composition of the bulb. Quartz is used because it permits higher operating temperatures than common glass. Quartz also provides high optical clarity and low expansion. The halogen or iodine part of the name refers to the kind of gas inside the bulb. Halogen is a family of gaseous elements, including fluorine, chlorine and, of course, iodine.

Halogen gasses are used in electrical lamps because of the phenomenon known as the *halogen cycle*. In this cycle, tungsten vapor given off by the glowing tungsten filament is redeposited back onto the filament, extending its life significantly. This also practically eliminates bulb blackening.

In the halogen atmosphere, the tungsten filament can carry a higher current for a given filament size, which produces more light. The light is also hotter and whiter. Photographers refer to this characteristic of light as *color temperature*.

To understand how hotter light is also whiter, you must understand a scientific analogy called the *black-body radiator*. Imagine a block of material that's absolutely black when it's cold. Then imagine that the block can be heated by a method that won't consume the material, no matter how hot it gets. With just a little heat applied, the block will take on a dull red color. With more heat, it goes through stages of cherry red, orange, yellow and finally white. The sun is a prime example of an object that's glowing with such intensity that its light output is nearly white.

In a halogen lamp, the tungsten filament glows much brighter and hotter than it could in a conventional bulb, so the light output is noticeably whiter. If you park two cars side-by-side at night, you'll see the difference in the color of the light between the one with conventional lamps and the one with halogen lamps. Although halogen lamps draw slightly more current than conventional lamps, you don't have to change wiring or fuses to substitute halogen lamps for conventional ones.

Headlamps containing two filaments carry the number 2 formed into the glass of the front lens at the top edge. There are three terminals on the back—one for the common ground, one for the low-beam filament, and one for the high-beam filament.

Before 1940, when "modern" headlamp regulations began, most cars had bulb-and-reflector headlamps. In this design, the body of the lamp contained a polished reflector and provision for installing a separate bulb. This was replaced by the *sealed beam* lamp. It has a sealed-glass body containing non-serviceable filaments and reflector. The non-sealed type was not optically inferior to the "better" sealed-beam lamp, except

In typical bucket installation of headlamp, chromed headlamp bucket is designed to accomodate a standard two-beam headlamp bulb.

Code number molded into glass at upper edge of headlamp (arrows) indicates number of filaments. High-beam-only lamps will have a one; high-low beam lamps will have a two.

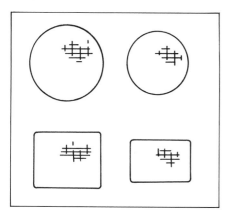

Headlamps are standardized into four versions. Large rectangular and the large circular lamps have both low and high beams in one bulb. Small rectangular and small circular lamps have either low and high or just high beams. These are used in 4-lamp systems.

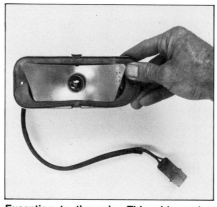

Exception to the rule: This side-marker lamp uses a parabolic reflector—most side-markers don't. Photo by Jim Barrett.

This OEM fog lamp is more decorative than functional. Although mounted close to road according to good practice, output is weak. And I discovered that lamps on both sides were poorly aimed when car was delivered.

that it required the owner to periodically check for and remove tarnish that inevitably built up on the reflector.

In 1955, a modest candlepower increase was permitted. Finally, in 1957, the first significant headlamp change occurred. That was the year the four-lamp, or *quad-lamp,* system was introduced, initially only on selected luxury cars. It soon spread throughout the industry. In this system, one lamp on each side of the car contains both low and high beams, while the other pair contains only the high beam. Although this system permitted optimum use of high beams, it did so at the further penalty of the low beams. In the quad system, the lamps have a smaller diameter than those in the two-lamp system.

Lamps containing both filaments use the same number 2 designation and three-terminal connectors as the older ones. The high-beam-only lamp contains one filament, so it carries the number 1 on the front lens, and it has a two-terminal connector. One terminal is for ground and the other for current feed. The high-beam-only lamp is illuminated when the low/high lamp is switched to high beam.

By the mid-70s, still another variation of lamps appeared. Rectangular quad lamps became available, the reason being to accommodate a lower hoodline. Then, in 1978, another system was introduced. This time it was a rectangular single-lamp system.

One thing to note about all four systems mentioned is that the maximum *candlepower* (light output) permitted is the same. This is based on laws established in 1940, when sealed-beam lamps installed on new cars became required by law. Sealed-beam and candlepower regulations vary from state to state. It's up to you to determine and comply with the laws of your state.

Lamp Reflectors—Many of the small lamps used in a car require no reflector. These include instrument-panel lamps and side-marker lamps. But lamps with higher output usually have a reflector to concentrate the light output for better efficiency.

Headlamps are the only ones that have the reflector built into the bulb. Other lamps use a reflector in the lamp-assembly body. These include taillamps and driving lamps, both of which have a replaceable bulb inserted into a weather-sealed housing with an integral reflector.

The purpose of any reflector is to capture all the light rays that radiate in random directions from the glowing filament and direct them into a single, concentrated beam. In precision optical systems such as headlamps and driving lamps, the reflector is parabolic in cross-section with the filament positioned at the *focal point* of the parabola. The parabolic reflector is the most efficient shape for maximum light output. In lamps where precise optics aren't criti-

cal, the reflector generally takes a more flattened shape to distribute light over a greater area. A car's dome lamp and courtesy lamps are examples.

Lamp Applications—The preceding sections described several kinds of lamps in order to explain various characteristics of lamps in general. This section discusses the various kinds of lamps according to *application*. This will help you select the specific kinds of lamps required for your project car.

Of all the lamps on the outside of a car, *headlamps* contribute the most to safe driving at night. Their task is to illuminate the roadway adequately for good vision, but without blinding oncoming drivers. For this reason, headlamps have two illumination levels, or *beams*. The *low beam* is satisfactory for low-speed, around-town driving, when following another car, and for temporary use on the highway when meeting an oncoming car. The *high beam* is suitable for all night-driving conditions where there are no oncoming cars and no cars immediately in front of you.

Because it's not optically possible to place both the low- and high-beam filaments at the exact focal point in a bulb, lamp designers compromise. The high beam is the critical one, so the low-beam filament is the one that's displaced. A *dimmer switch* is wired to permit the illumination of only one filament at a time.

Driving lamps and fog lamps are used by the discerning driver to supplement the OEM headlamps. Few new car or truck models come equipped with driving lamps—even as an option—although some manufacturers offer fog lamps on selected car models. I suspect that they're used more as a styling or image gimmick than for any practical use. The fog lamps that came with my personal car are woefully inadequate. However, the aftermarket offers a wide range of driving lamps and fog lamps for the serious driver.

In street driving, driving lamps are intended to supplement the headlamps under two specific conditions. The most obvious one is high-speed driving at night on the open road with minimal oncoming traffic, where the lights won't blind oncoming drivers.

The second use is to supplement the headlamps when the low beams don't provide enough light, but the high beams provide too much. Driving lamps for this application must be carefully aimed to avoid blinding oncoming drivers.

Driving lamps for off-road vehicles are another matter, as described on page 148 and in above photo.

Fog lamps serve two purposes. The obvious one is to illuminate the road under foggy conditions, when the use of conventional headlamps would create a dazzling effect. In this application, the fog lamps are wired to operate in both the parking-lamp and low-beam modes.

The second use is to illuminate the edges of the road, even when there's no fog. In this application, fog lamps can be used with both low and high beams. Check local ordinances.

When you wire fog lamps and driving lamps into the headlamp circuit, use relays to control them. Relays are more expensive, and more complicated to wire, but they reduce the incidence of blown fuses and eliminate the possibility of having the wrong lamp on at the wrong time. This advice is based on personal experience—I was once stopped by a polite but firm officer who took a dim view of being accidentally blinded by my driving lamps! The appropriate diagram for wiring these lamps is on page 33.

Off-road racers rely on high-output driving and fog lamps to illuminate as much as the terrain as possible. Setup shown here is basic—two headlamps and two driving lamps. Many off-road setups use more lamps than this—mounted in locations to minimize damage during racing—which likely would be illegal for street use. Lens covers protect lamps when not in use—they're also good advertising for lamp manufacturer! Photo by Ron Sessions.

MAGNETIC LOADS

Magnetic loads rely on the electrical principle that a magnetic field is generated around any electrical conductor. If you wind the conducting wire into a coil, the magnetic field becomes concentrated. And if you put a metal core inside the coil, the magnetic field is even further concentrated and can accomplish work. You may remember that this is also the basic principle behind relays and solenoids, discussed earlier.

In all cars, two important kinds of loads depend on a concentrated magnetic field. These are the *ignition system* and the *electric motor*.

Ignition System—The type of ignition system you'll use in your project car depends on two basic factors. One is the type of equipment the factory provided. The other is the selection of replacement equipment available in the aftermarket. Because of the ignition system's unique role as an engine component, this section covers only the basic operating principles as they relate to the car's electrical system. If you want a specialty system, you'll need to consult a local speed shop for details. Also, pick up a copy of HPBooks' High-Performance Ignition Handbook.

An ignition system transforms low voltage from the battery to the high voltage needed to jump a sparkplug gap in the sparkplugs. The high voltage is routed—*distributed*—to the appropriate sparkplug at the exact moment it's needed in the combustion cycle.

The rate at which sparks are delivered by the ignition system depends on how fast the engine is running and how many cylinders the engine has. Spark rate can be computed by using the following formula: *Revolutions per minute (rpm)* $\div$ *2 X number of cylinders = sparks per minute.* Example: *For a 4-cylinder engine running at 5,000 rpm, we get: 5,000 $\div$ 2 X 4 = 10,000 sparks per minute.*

It's evident that an ignition system produces sparks at an enormously fast rate. To understand how 12 volts becomes 25,000 volts, and does so 10,000 times a minute, let's examine how one spark is generated.

Coil Theory—An ignition coil works on two basic principles. The first is the magnetic-field principle just discussed. The second is the *induction* principle. As described on page 18, moving a conductor through a magnetic field induces current into the conductor. However, both phenomena—magnetism and induction—occur inside the ignition coil without need for any moving parts.

The ignition coil consists of a *primary coil* and a *secondary coil*. Shutting off current flow in the primary (low voltage) coil causes its magnetic field to *collapse*.

Ignition coils like this one have been used on cars for decades. Newer electronic systems incorporate coils of different shapes, but basic function is the same—to create a high-intensity spark.

VOLTAGE INDUCTION IN COIL

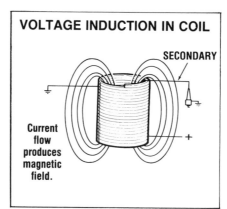

When current flows through primary winding of ignition coil, a magnetic field is created. Drawing courtesy of Chrysler Corp.

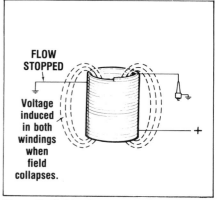

Magnetic field collapses and induces high voltage into secondary winding when primary circuit is opened. Also note that some voltage is induced into primary, but it is of no value in creating spark. Drawing courtesy of Chrysler Corp.

The motion of the collapsing field past the secondary (high-voltage) coil causes current to be induced into it.

Another electrical phenomenon, called *electrical transformation*, also takes place in a coil. The primary coil has only a few hundred turns of wire, while the secondary coil has several thousand. Electrical transformation determines that the ratio of *original current* to *induced current* is the same as the ratio of primary-coil turns to secondary-coil turns. So, the ignition coil actually multiplies battery voltage of about 12 volts to 25,000 volts or more each time the primary coil is energized and then turned off. The resulting voltage is routed through the distributor to the appropriate sparkplug, where it ignites the air/fuel mixture in the combustion chamber. Current is reduced in the same proportion.

The coil's primary winding has a positive (+) terminal and a negative (−) terminal. The positive terminal is supplied from the ignition switch; the negative terminal is connected to the distributor points, where the primary circuit is opened and closed. The secondary coil winding is attached to the center tower of the coil, which leads to the center tower of the distributor cap. A *rotor* driven by the distributor shaft transfers current to the appropriate sparkplug wire by bridging the gap between the cap's center tower and the tower corresponding to the cylinder to be fired. The current races down the sparkplug wire to the spark-

plug, where a spark occurs that ignites the air/fuel mixture in the cylinder.

Points & Condenser Ignition—The points are a set of electrical contacts that open and close the primary circuit. The points are positioned between the coil and ground. A small cam with one lobe per cylinder operates the points in coordination with other engine functions at one-half engine speed. The amount of time the points remain closed is called *dwell* and is read in degrees of crankshaft rotation. The longer the dwell, the more time there is for the primary field to build up.

COIL DETAIL

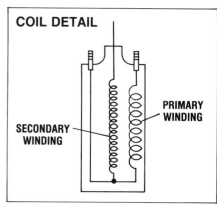

This drawing shows what's inside a typical ignition coil. Primary winding contains relatively few windings; while secondary has considerably more.

CONDENSOR DETAIL

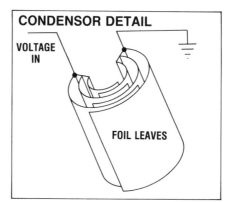

Condenser consists of two sets of foil leaves, one connected to source of voltage spikes and the other to ground. Not shown are insulators between the leaves. Condensers are wrapped tightly into a cylinder for compactness. Metal case acts as ground connection.

Ignition coil has two small terminals and one large one. Small terminal at left (BAT or +) connects to ignition voltage source; terminal at right (DIST or −) connects to distributor points. If car has a tach, it gets signal from this terminal, as well. Large center post connects to distributor cap.

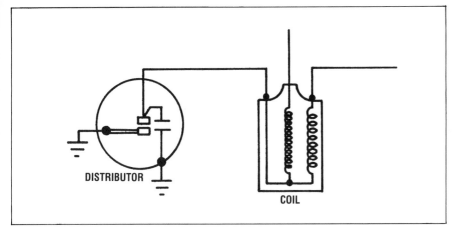

Diagram shows how voltage applied to coil input is grounded when distributor points close. When the points open, field in coil collapses. Voltage spikes are dissipated by condenser.

turbs an electrical field in the sensor, indicating the need for a spark to occur. Some aftermarket units use a light source and a photoelectric cell with a shutter wheel to chop up the light beam, each light segment corresponding to one cylinder's ignition. The most accurate ignition-triggering mechanisms use either system, but are mounted at the crankshaft damper. Almost without exception, such systems are restricted to race-car use.

With the advent of on-board computers, electronic-ignition control has moved into the space age. However, such computer systems are available only as factory equipment on new cars, and are not yet available in the aftermarket. Also, because of their complex nature, they are not recommended for do-it-yourself electrical systems. I suggest using a points-and-condenser ignition or an aftermarket electronic ignition.

Resistor & Bypass Circuit—Despite the buffering action of the condenser, ignition points may wear out prematurely if they carry 12 volts routinely. The life of a coil is also determined by the amount of current flowing through it.

Many OEM ignition systems on American cars include a *series resistor*

At the instant before the points open, current is racing through the primary coil in search of ground. When the circuit opens, induction starts, as just outlined. Most of the current is induced into the secondary coil, but about 250 volts is induced into the primary coil and has nowhere to go. Its most likely path would be to jump the gap between the now separating points. At the rate of 10,000 jumps per minute, it wouldn't take long for the points to burn up. So, the *condenser* provides an alternate path for the primary induced voltage.

Inside the condenser are two sets of foil leaves separated from each other by an insulator. One set of leaves is connected to the condenser lead, and the other to ground through the condenser's metal case. At the instant the points open, the condenser "looks" like a ground to the rushing electrons. By the time they pack into the condenser, they "realize" the condenser isn't really a ground, so they rush back into the primary coil where they contribute to the induced voltage in the secondary coil.

Electronic Ignition—At the expense of oversimplification, I think of the *electronic-ignition module* as nothing more than a substitute for ignition points and a condenser. An electronic-ignition system still requires a coil, though the coil's shape may vary somewhat from those used in conventional ignition systems.

Essentially, the module contains a transistor that opens and closes the coil's

primary circuit just like points. The module depends on a signal from the distributor to indicate when a spark should occur. Early units actually used points for this. However, the points had an extended life because they weren't carrying heavy primary voltage.

Modern electronic ignitions rely on a pickup sensor that reacts to teeth on a trigger wheel, one tooth per cylinder. The presence of the passing tooth dis-

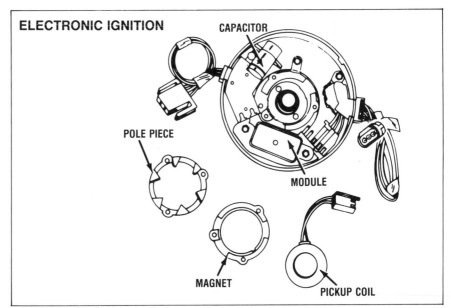

ELECTRONIC IGNITION

In this electronic-ignition distributor, points have been replaced by electronic components. Relative movement between teeth on pole piece and magnet create signals detected by pickup coil to determine when to "fire" ignition coil. Drawing courtesy of Chevrolet Motor Division.

Cars with point-type ignition often rely on a resistor to prolong point life. Battery voltage (from right hand wire) is cut by approximately one half. Bypass wire from the starter solenoid is attached to the distributor side of resistor (left) to provide full voltage during engine cranking.

This is where other end of resistor bypass wire is attached. During engine cranking, full battery voltage is available at "I" terminal. If your car has an ignition resistor, use starter solenoid with bypass terminal—identical solenoids are made *without* "I" terminal.

that reduces current flow to the coil by approximately half. The resistor may be built into the coil, mounted to the firewall, or built into the harness in the form of a calibrated-resistance wire.

During engine start-up, a full-strength spark is desirable. So, a *bypass circuit* is included in the starter circuit to provide full battery current to the coil only while the starter motor is operating.

In the era before electronic ignition, starter solenoids had an extra terminal labeled I for ignition or R for resistance bypass. The terminal was wired directly to the coil's + terminal to bypass the built-in resistance while the solenoid was energized. As soon as start-up was completed, the solenoid de-energized and the bypass was turned off.

A final word on ignition systems: Electronic tachometers rely on the coil's on-off current for an operating signal. The sender wire is usually connected to the coil's negative terminal. For more on electronic tachometers, see page 44.

Electric Motors—Because electric motors are so commonplace, many people give little thought to the fact that their operation seems to defy simple logic. In fact, according to the Delco Remy Division of GM, an electric motor is an *educated short circuit*. And, unlike a gasoline engine, an electric motor develops

maximum torque when *stalled*. The reasons for these seemingly impossible facts should become clear as you read the following discussion on the operating principles of the electric motor.

Motor Principles—An understanding of the electric motor begins with the fact that a current-carrying electrical conductor has a magnetic field around it. Place this conductor between the poles of a horseshoe magnet and there'll be two magnetic fields—one from the conductor and one from the magnet.

In any magnet, including a conductor, the lines of magnetic force move from the north (positive or +) pole to the south (negative or −) pole. The lines of force in the wire conductor obey the *right-hand rule of magnetism*. The rule works like this: If you grasp the conductor in your right hand with your thumb pointing in the direction of current flow, the direction of your fingers will indicate the direction of the magnetic field.

Let's suppose a horseshoe magnet is positioned so the north pole is at the bottom, meaning the magnetic lines move upward toward the south pole. The field in the conductor is moving clockwise. On the left side of the conductor, both magnetic fields move the same way, so they *reinforce* each other. On the right side, they're trying to move in opposite

directions so they somewhat *cancel* each other. With a strong, reinforced field on one side and a weakened one on the other, the conductor is forced to the right. This is the basic magnetic principle used to turn an electric motor.

Now, imagine the wire conductor bent into a loop and positioned between the two poles of an electromagnet. Further imagine that each end of the loop terminates in a semicircular *commutator*. A pair of *brushes* riding against the commutator are connected to a battery. Current flows through the electromagnetic coils, through one brush and commutator, through the loop of wire, through the second brush and commutator and back to the battery.

Several magnets are instantly created by the flowing current. The electromagnet forms the two poles corresponding to those of the horseshoe magnet. And a field is created around each leg of the wire loop.

The interaction of the magnetic fields produces a rotational force in the loop. When the loop has made 1/2 revolution, each commutator is in contact with the opposite brush from the starting position. A new field is created in the loop and the cycle repeats. The rotating loop is called the *armature* of the motor. Imagine multiple wire loops, each connected to a pair

HOW A MOTOR WORKS

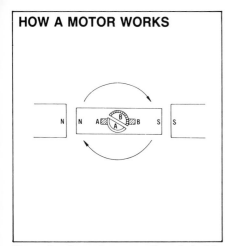

These drawings show operation of electric motor during one revolution. Note permanent magnets at left and right. Current flows from brush A, through commutator A, through armature coil, through commutator B to brush B. Armature becomes an electromagnet. Initially, armature poles are *repelled* by like poles of permanent magnet, starting clockwise rotation.

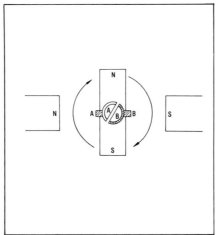

When armature has turned 90°, North pole of armature is attracted toward South pole at right. South pole of armature is attracted toward North pole at left. Rotation continues.

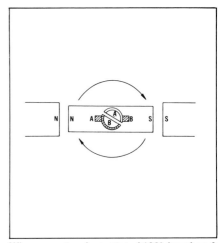

When armature has rotated 180°, brushes A and B are now contacting commutator segments B and A, so current through armature reverses, reversing armature polarity. Unlike poles repel and rotation continues.

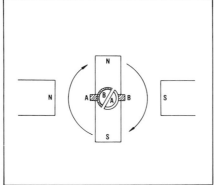

When armature has rotated 270°, unlike poles begin to attract, and rotation continues.

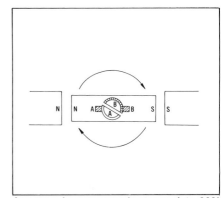

Armature has now made a complete 360° revolution and is back where it started. Again, polarity reverses, unlike poles *repel* and cycle repeats. Drawings courtesy of Chrysler Corp.

of commutator segments, and you'll have an idea what a real electric motor is like.

Here's a point you may have missed. A motor has a loop of wire rotating in a magnetic field—the exact conditions necessary to make a generator! So, could it be possible for the motor to try to generate current at the same time it's running? You bet it is.

The current generated within a running motor is called *counterelectromotive force* (CEMF). The strength of the CEMF is directly proportional to motor speed. As the motor runs faster, more CEMF is generated, causing less battery current to flow. Although CEMF may approach battery voltage, it never equals it. Otherwise the motor would eventually stop. In fact, the top speed of a motor is limited by the CEMF. So, an electric motor is self-limiting at top speed. CEMF is also the reason an electric motor acts like an educated short circuit. Before current flows in a motor, the internal resistance may be as low as 1/100 ohm—definitely a short circuit to ground! But as soon as current flows, and the motor starts to turn, CEMF automatically regulates current flow, so the expected short circuit never occurs.

The turning force of the wire loop,

SIMPLE MOTOR

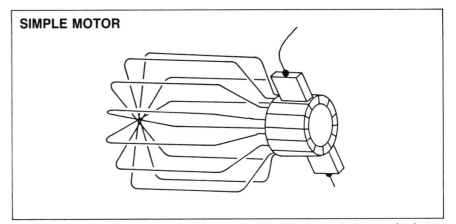

Simplified drawing of armature shows how each wire loop is attached to a pair of commutator bars. Commutator bars are contacted by a pair of brushes. As armature turns, commutator bars are continually moving past brushes, making each loop into a momentary electromagnet.

STARTER MOTOR

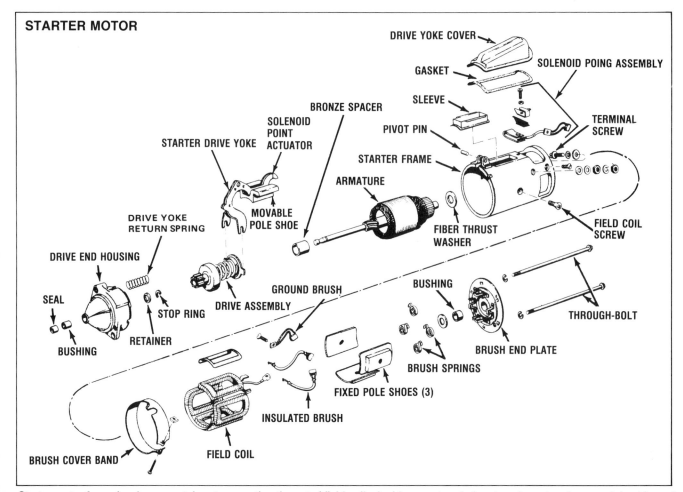

Starter motor frame is a heavy metal part supporting the set of field coils. Inside, armature is free to spin on bearings contained in end housings. Brushes are contained in brush holders in end housing. Drawing courtesy of American Motors Corp.

called *torque,* is directly proportional to current flow. That's why a motor puts out its maximum torque when its speed is lowest, when CEMF is lowest. An electric motor operating against a load that drags down its operating speed causes the motor to overheat. The heat is caused by high current flow at work. The current is relatively unresisted because at low motor speeds there's little CEMF generated.

Perhaps now you can see why an electric motor can be called an "educated short circuit." To summarize, when current is first applied to the motor, it can flow almost entirely unimpeded. Torque, which is proportional to current flow, is also high. As motor speed increases—it does so rapidly—CEMF increases, automatically limiting current to a safe rate.

Motor Components—Most of the components have already been described in the discussion on how a motor works, although in basic form. These and the remaining ones are described in detail here.

The *motor housing* is a heavy iron frame to which the *field-coil cores,* called *pole shoes,* are attached. The *field coils* are wrapped around the pole shoes and are insulated by a coating on the wire. The wire loops described earlier are contained in the *armature,* a kind of axle shaft that rotates within the motor housing. Each loop terminates in a pair of *commutator segments.* These metal segments are insulated from each other and from the armature shaft. Thin sheets of *mica* are often used as the insulating material.

The armature is supported by *bearings*

or *bushings* assembled into the motor housing. Finally, a pair of insulated *brushes,* held in insulated pockets within the housing, are positioned to rub against the commutator bars. *Terminals* are wired to carry current to the field coils and brushes.

Electric-Motor Types—Large, powerful electric motors, such as starting motors, use electromagnetic field coils. Smaller motors, such as those used for windshield wipers, power windows and the like, may use a permanent magnet in place of the field coils. The permanent-magnet motor is less expensive to build, and is more than adequate in low-demand applications.

Power output is a limitation of the permanent-magnet motor. Size-for-size, its output is not as great as that of a field-coil motor. In OEM applications,

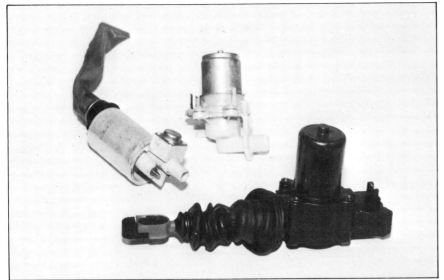

Electric motors come in all sizes, shapes and applications. In these examples, the motor is the cylindrical shape. From left to right are: in-tank electric fuel pump and filter, windshield-washer pump and electric door-lock actuator. Corrugated part on door-lock actuator is the lock drive rod. Motors are permanent-magnet type. Door lock actuator is reversible.

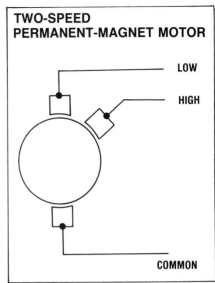

TWO-SPEED PERMANENT-MAGNET MOTOR

Third brush can be added to permanent-magnet motor to make it operate at two speeds. When low-speed and common brushes at 180° apart are used, motor runs at slow speed. When high-speed brush and common brush are used, motor runs faster. If high-speed brush could be moved farther clockwise, motor would eventually begin to slow down and stop.

permanent-magnet motors generally operate through a torque-multiplying gear train. You'll often find the higher-powered field-coil motors attached to a gear train as well. This may be more for slowing down the action than for torque multiplication, though.

If you decide to install a custom motorized feature in your car, be aware of the output limitations of the permanent-magnet motor. Better safe than sorry.

The available speeds of *multi-speed motors* may be provided by one of three common methods: *resistors, shunt windings* or *multiple brushes*.

A typical *resistance-controlled* motor is the heater blower. The control switch selects from a variety of resistors that permit different amounts of current to flow. Remember the speed and power of a motor are governed by the strength of

the armature magnetic field, created by current flow. Reducing current flow weakens the field, slowing the motor.

Motors with a *shunt coil* have two windings in the field coil—the usual one and a much smaller one threaded throughout. With the main coil energized, the motor runs fast. With the shunt also energized, the main field is weakened, slowing the motor. This type of circuit allows two speeds, and is used in many GM wiper motors.

The speed of a permanent-magnet motor can be controlled by a set of *multiple brushes*. As mentioned earlier, the common location for brushes is 180° opposite each other. If you could hold one brush stationary and somehow rotate the other one in the direction opposite to armature rotation, the effect would be something like advancing the ignition on

an engine. The motor would run faster—up to a point. If you rotate the brush too far, the motor will slow down and finally stop.

It's not practical to make a motor with a movable brush—although it's been tried—but it is practical to add a third brush in an advanced position. By selecting this brush with the control switch, you can make the motor run faster. This is a common method of achieving two speeds with a permanent-magnet motor.

I bought this set of terminals and electrical supplies from DRI Industries. It contained all the kinds of terminals needed to wire my car, although I had to buy extra quantities of some. As an added bonus, they offered a wire crimper at a discounted price, along with the terminal assortment.

This chapter covers the minor components that don't fit well into the categories discussed in previous chapters. Many are items you'll be using by the dozen, anyway.

FUSE PANELS

To be absolutely truthful, you don't need a fuse panel in a car's electrical system. Instead, you can use an in-line fuse holder in each circuit. However, using these holders instead of a fuse panel is impractical—assembly is difficult and they are not always convenient when a fuse needs to be replaced. Fuse panels make wiring and fuse replacement much easier.

There's a wide variety of fuse panels to choose from, depending on the "donor" car you're taking your electrical components from, and whether you want to use glass-capsule fuses, Autofuse-type

fuses or ceramic fuses. Small panels for all three are available in the aftermarket. Or, you can remove a serviceable panel from a donor car and adapt it to your car.

On page 36, an argument is presented both for and against the use of circuit breakers versus fuses. If you decide to use fuses, you'll need an appropriate fuse panel.

Selecting an Aftermarket Fuse Panel—There are only a few things to consider when choosing an aftermarket panel. First, of course, is the type of fuses you'll be using. Then, you must make sure the panel will fit into the space available. Many aftermarket manufacturers make fuse panels of varying lengths, so you may need to buy a variety of sizes to accommodate the space and fuse requirements of your car.

Many aftermarket fuse panels also include—or are a part of—another kind

of component called a *terminal block*. This component is used to connect major wiring harnesses. Terminal blocks are usually made of an insulating fiberboard with the electrical conductors riveted in place.

Frankly, I haven't seen too many general-purpose aftermarket fuse panels that have impressed me much. I suggest that if you've chosen fuses over circuit breakers, go the rest of the way and adapt a used, factory-type fuse panel, as described below.

Adapting a Used Fuse Panel—If you remove and adapt a fuse panel from an existing car or truck, there are a few things to consider:

First, if you're building a kit car, the manufacturer may make specific recommendations. Check to determine why a particular panel is required. It may be that you'll be taking large enough portions of the harness from the donor car and the fuse panel is an integral part of it.

If the decision is up to you, be sure the panel will accommodate the style and number of fuses you want, making allowance for adding circuits at a later date. You may need to refer to chapters 10 and 11 to determine this.

Next, be sure the panel you've chosen will fit the allocated place in the car. Measure if you're not sure.

Finally, check to be sure the vehicle wasn't involved in a fire and that it didn't suffer massive electrical damage. Charred wiring is a definite sign that you're looking in the wrong car! Then disconnect the battery, if the donor car has one. Remove the screws or bolts holding the fuse panel in place, then pull as much of the existing harness loose from the under-dash area as you can. Finally, cut wires free from the harness, leaving at least 1 foot of wire attached to each cavity in the fuse panel. The wire ends will be spliced to the new harness when you install it, so leave as much wire length as possible.

TERMINAL BLOCKS

Before discussing terminal blocks, I explain what bulkhead and harness connectors are used for in factory installations. This may help you see why they're not needed in your project car.

When a car goes down the assembly line, wiring is installed in sections called *harnesses*. There's a harness for the instrument panel, one for the front lighting, one or more for the engine, one for rear wiring, and perhaps several more for optional equipment.

Each harness is designed to go into the car at specific points of assembly, usually before upholstery, dash-panel cover and other interfering parts are installed. This means that individual harnesses eventually must be connected to each other to form the completed electrical system. Various harness connectors make this a relatively simple, hard-to-mess-up job. At the point where the harnesses pass through the firewall, or bulkhead, the outside harness plugs into the inside harness by means of a huge *bulkhead connector* may have up to 40 or 50 cavities.

Terminal blocks aren't used often in

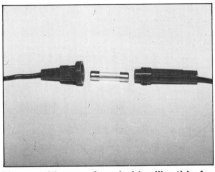
You *could* use a fuse holder like this for every circuit in your car: Jeep Corporation used to do just that! But modern thinking is to put all your fuses into one panel.

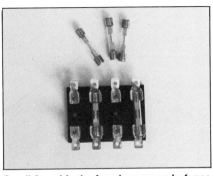
Small fuse blocks for glass-capsule fuses come in various sizes and shapes, can be assembled into larger panels. Photo by Jim Barrett.

GM uses tip-down fuse panel in most of its cars. These are convenient to service and simply push up and lock when you're done.

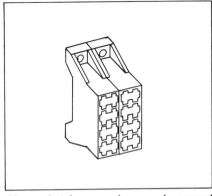
Aftermarket fuse panels come in several configurations. This one accomodates blade-type Autofuse. Several units can be stacked as shown. I bought mine by mail order from Ron Francis Wire Works.

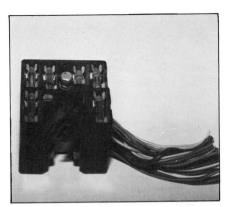
If you plan to salvage fuse panel from a donor car, leave as much wire attached as possible—1 foot is probably a desirable minimum. Wires on this one are cut shorter than that but could be made to work.

Production cars rely on a bulkhead connector to attach harnesses outside of body to those inside. Because many wires are crowded into a fairly small space, these are sometimes called "high density" connectors.

Fuse panel and main wiring harness from my Pinto donor car. Panel mounts under dash. Engine harness (left) passes through bulkhead and plugs into back of panel.

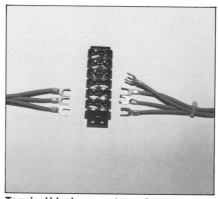

Terminal block connects one harness to another. Each screw in left column is electrically joined to adjacent screw in right column. A barrier insulates each pair of matching screws from pairs on either side.

Starting at bottom, first wire is connected to the one directly opposite. So is second wire. Third left wire connects to third and fourth right wires, which permits operation of two devices from one voltage source.

factory wiring jobs. Auto manufacturers rely more on specialized bulkhead connectors and harness connectors as those described. Such connectors are custom-designed for a specific car and are effective in reducing incorrect wire connections on an assembly line. But they're not easily adapted to a project car, nor are they needed.

The way you wire your project car will be entirely different. You can install the wires at your convenience, rather than at the demand of an incessantly moving assembly line. You can take the time to install the wires one at a time, or you can construct partial harnesses similar to the ones used in the factory, to be installed and eventually hooked together. Chapter 11 describes two harness-construction methods. For now, let's assume that you have decided to make two or three harnesses—one for the car's passenger compartment, another one for under the hood, and perhaps a third one for the instrument panel. How do you connect the various harnesses? You use terminal blocks.

A terminal block is usually a plastic or resin casting with two rows of screw heads showing. Each screw is separated from its neighbors in the row by small partitions. But, each screw in one row is electrically connected by a hidden *bus bar* to the screw directly opposite it in the other row.

Suppose your under-hood harness contains eight wires that must connect to a matching set of eight wires in the under-

dash harness. This is a simplified example, because there are usually far more than eight wires in an under-hood harness. When you install the under-dash harness, you'll run the eight wires through a hole in the firewall and attach each to a screw in one row on the terminal block. To complete the job, attach the eight under-hood wires to the screws in the other row, making sure that each pair of wires is attached to matched screws. Ring terminals are commonly used to connect wire ends to terminal-block screws.

What you've just done is connect two harnesses together without using solder or solderless butt connectors. Both solder and solderless connectors *would* work, but they're inconvenient to disconnect readily for later service or diagnosis work. In addition, terminal blocks make it easy to change wires around if you make a mistake or if you want to add another branch to a particular circuit. As with fuse panels, it's a good idea to choose a terminal block with a few extra terminals for possible future circuits.

FLASHERS

Generally, there are two kinds of flasher units—those for *turn signals* and those for *hazard flashers*. Although they are similar in appearance—often differing only in color—and they have identical terminals, turn-signal flashers and hazard flashers are usually not interchangeable. An exception is when trailer lamps are added to the circuit, as

described on the facing page.

Turn Flashers—These units are what make the front and rear turn signals and dashboard indicators flash on and off. All factory cars built after 1967 also have side-marker lamps. In most of these, the front markers also flash.

The turn flasher has a set of movable contact points and a heating coil operated by current flowing through the unit. During turn-signal operation, current flows through the points, then the heater, the turn-signal switch and finally the lamps. Front and rear turn lamps selected by the turn-signal lever turn on. When the heater warms up, it temporarily opens the points, and the turn-signal lamps go off. As soon as the heater cools, the points close and the cycle repeats.

If one exterior turn-signal lamp is burned out, there's not enough current flow in the circuit to make the heater work, so the remaining lamp simply comes on and stays on. So do the turn-signal indicators on the dashboard. This is a safety factor intended to alert you that the turn-signal lamp is burned out. Compare this to the operation of the hazard flasher, below.

Hazard Flasher—The hazard flasher is intended to operate all four turn-signal lamps simultaneously. Because the flasher is an emergency device, it's designed to operate even if a turn-signal lamp is burned out. In fact, it will operate with only one of the four lamps functioning.

The hazard flasher also depends on a

Turn-signal flashers and hazard flashers look alike and have identical terminals. But they are not intended to be interchanged unless you understand the consequences, explained in text below.

TURN FLASHER

CONTACTS CLOSED   CONTACTS OPEN

Turn-signal flasher is designed to operate a predetermined number of lamps. Current flows through lamps and heater coil at same time. The flash rate depends on load created by lamps.

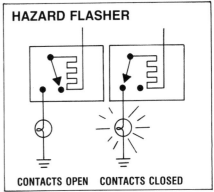

HAZARD FLASHER

CONTACTS OPEN   CONTACTS CLOSED

Hazard flasher is not dependent on number of lamps in circuit. Current flows to lamps, then to heater, in turn.

set of movable points and heating element, but the configuration is entirely different from the turn-signal flasher. In the hazard flasher, the points are normally open, so when you first turn on the flasher, current flows through the heater and lamps. Because of the high voltage drop across the heater, the lamps do not light. As the heater warms it closes the points, which shunt all current directly to the lamps, lighting them. With no current flowing through the heater, it cools and reopens the points. Then the cycle repeats.

Interchanging Turn Flashers & Hazard Flashers—Ordinarily, you don't. Although the terminals are identical, and both units will make the turn signals flash, you shouldn't use a hazard flasher in a turn-signal circuit. It defeats the safety-design intent of warning you of a burned-out lamp.

There is one exception, though. When you're pulling a trailer, the added load of trailer lamps on the turn-signal circuit is best handled by a hazard-flasher unit. They are also sold on the market as *heavy-duty flashers*, which is another name for the same device. Just be sure to reinstall the regular turn-signal flasher after you're done pulling the trailer.

SOLDERLESS CONNECTORS

Solderless connectors are used to make permanent junctions. They fall into three general categories: *solderless terminals, solderless butt joints* and *solderless closed-end connectors*. Within these

categories, you'll find both insulated and non-insulated connectors. Connectors discussed here are general purpose, as compared with the specialized connectors used in factory harnesses.

All solderless connectors have one important feature in common. They can be permanently attached to a wire without use of solder. And, they are neither removable nor reusable. The easiest way to remove a solderless connector from a wire is to cut the wire as close as possible to the connector.

There's one additional, and unique, kind of solderless connector described at the end of this section. It's a removable *snap splice connector* intended to join a wire to one that's already in use. These are often used to add trailer wiring to a car's electrical system.

Solderless Terminals—These are used at the ends of wires that are attached to components or other wires. Insulated and non-insulated types are available. Insulation is either *full-collar* (entire terminal) or *crimp-area* (barrel only).

Solderless terminals fall into three general categories, depending on the shape of the terminal's "business end."

Ring terminals are used to attach a wire to a threaded stud on a component. Common applications include terminal-block screws and studs on Ford-type starter solenoids. They are also used with screws or bolts to attach ground wires to a car's metal body or chassis.

An advantage of the ring terminal is that it cannot disconnect itself from the

screw or stud, even if the fastener loosens slightly—the screw or nut must be removed. In some instances, this could be a slight disadvantage if you need to disconnect a terminal.

Due to the nature of their installation procedure, ring terminals do not come with full insulating collars. Once installed, the ring portion of the terminal cannot come into contact with anything but the the stud or screw it's attached to. They are, however, available with insulated crimps.

Spade terminals are also used on threaded studs or screws. The advantage a spade terminal has over a ring terminal is that it is forked so it can be removed by loosening the fastener enough to free the legs of the terminal. However, this could also be a disadvantage because the wire may disconnect if the fastener becomes loose.

Spade terminals are also installed in such a way that they cannot be provided with insulating collars, but are available with or without crimp insulators.

Disconnect terminals come in both male and female versions. Depending on application, the female type may or may not need a full insulating collar. Both kinds are also available with no insulation and with crimp insulation only.

The *female disconnect terminal* is actually a type of socket, and it's usually on the *feed side* of a connection. Often, the female terminal is insulated by a built-in collar. This protects the terminal from shorting if it comes in contact with a

Neatness counts. Keep terminals organized and labeled to minimize errors when you're working under pressure. This assortment came with a full set of labels.

Small terminal kit is inexpensive, handy for quick repairs. Photo by Ron Sessions.

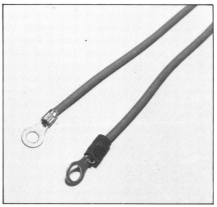

Use ring terminals to attach wires to components with threaded stud or screw terminals. Ring terminal at top is non-insulated for use with shrink wrap. One at bottom is insulated, and is suitable when using several terminals in close quarters.

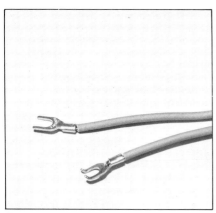

Spade terminals: One at top has conventional straight-legged fork, so it can be slipped under a slightly loosened screw—it can also slip out! Bent-tip terminal at bottom is less likely to do this—you can buy these prebent, or as I did, bend tips of conventional terminal with needle-nose pliers.

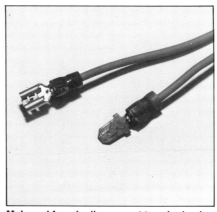

Male and female disconnect terminals plug together. Ones shown here have insulated barrels only; other types are fully insulated. When using disconnect terminals, make sure they have same blade width.

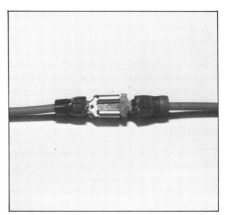

When connecting male and female terminals, be sure they're pushed all the way together.

grounded component or metal part.

Female disconnect terminals are most often used to attach a wire to the built-in male blade on a load component. A good example is the headlamp. The back side of the lamp has protruding male blades, while the wires that connect to the headlamp are fitted with female disconnect terminals.

The *male disconnect terminal* is intended to plug into a matching female disconnect terminal. It's usually used on the wire being fed; that is, it's not often found on the *feed side* because it could accidentally short if it touched anything grounded. Male solderless terminals

usually do not have built-in insulating collars, but they are available with or without crimp insulation.

Solderless Butt Connectors—These are used to join two wires end-to-end, or butted. Hence, the name. The butt connector is tubular in shape, with a full insulator covering it. The wires to be joined are inserted into opposite ends of the butt and are crimped in place. It's also possible to use a butt connector to join two wires to one, again depending on the size of the connector opening. However, a closed-end connecter is preferred for joining three or more wires.

Closed-End Connectors—These are

shaped something like an old-fashioned derby hat. The crown of the hat is large enough to accommodate several wires at once, while the brim of the hat is large enough to fit over the wire's insulation, ensuring an insulated connection. Other than their ability to accommodate a larger number of wires, closed-end connectors may be regarded as interchangeable with butt connectors. Like other solderless connectors, closed-end connectors are installed by crimping.

Snap-Splice Connectors—This is a reusable connector used to join two wires without need of cutting and splicing. It is commonly used to tap into existing wir-

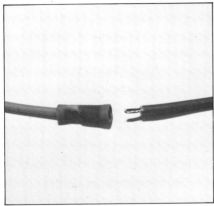

Butt connector is used to join wires end-to-end. Here, it has been crimped onto one wire; other wire is ready to be inserted. Note short amount of exposed conductor.

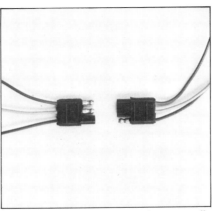

Multi-wire disconnect terminals are usually *polarity protected*—that is, they can be plugged together only one way. These molded-plastic connectors are usually sold with short lengths of wire attached. This one is fully insulated. See page 83 for more information. Photo by Jim Barrett.

Closed-end connectors for automotive use are installed by crimping. Don't confuse them with type used for house wiring. This one is used to join two wires—larger ones could be used to join three or more wires. Make sure connector insulation covers exposed conductors of all wires.

Female disconnect terminals are often used to attach wires to components. If component terminals are widely separated, no insulation is needed. But if area is at all crowded, use insulated terminals for added protection against short circuits.

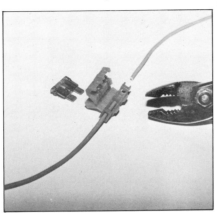

Snap-splice connectors are sized according to wire gage. Of all connectors mentioned, these are only ones that can be reused. One shown here has a built-in fuse. To install, insert wires into slots provided and follow directions in photo at right. It is not necessary to strip insulation.

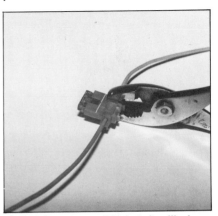

Use pliers to snap lid shut. It will pierce insulation and make contact without cutting conductor.

ing, such as when you add a trailer connector to a car. See page 138.

The snap-splice connector opens on a hinge. When it's open, you'll see a groove to accommodate the insulated original wire and another dead-end groove to accommodate the wire you're adding. To complete the connection, hold the two wires in place and close the lid. Then use a pair of pliers to squeeze the sliding metal contact into place. When the contact moves, it pierces the insulation of both wires simultaneously, providing *continuity* between them. A second flap snaps shut over the metal contact to protect it from shorts.

Although a snap-splice connector should be regarded as permanent, it can be removed and reused. Be sure there's no current in the wires before removing it. Open the flap to gain access to the hinged lid. Pry open the lid and remove the connector from the wires. Push the movable metal contact back to its original position. The connector is now ready for reuse. Be sure to tape carefully over the wire insulation that was pierced by the connector.

Battery-Cable Terminals—These are available in a number of configurations, determined by the type of battery being used. All can be divided into two basic

types: *clamp-* and *lug-type terminals.*

Clamp-type cable terminals are used on batteries with top posts. The terminal is designed to slip over the post, and is then tightened with a clamp bolt. Because the positive post on a battery has a slightly larger diameter than the negative post, high-quality clamp terminals are sized accordingly. Cheaper kinds are not.

Lug-type cable terminals are used on batteries with side terminals. They're also used at the other end of the battery cables where one cable attaches to the starter solenoid or starter motor, and the other to ground on the engine block or

BATTERY-CABLE TERMINALS

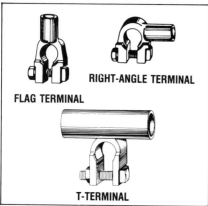

Cable terminals for top-post batteries include standard flag terminal (top left), right-angle terminal (top right), and T-type terminal (bottom) for connecting two or more batteries in series. Ones shown are installed by crimping.

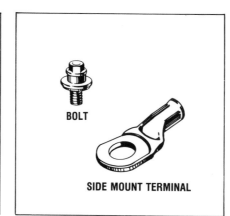

Cable terminal for side-mount batteries uses large ring terminal and bolt. This one is installed by crimping.

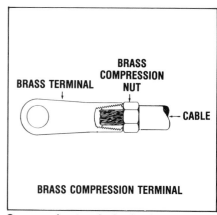

Compression terminals attach to cable by tightening brass compression nut.

Least desirable are temporary terminals that clamp to cable by tightening two bolts.

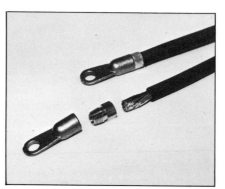

To install compression terminal, simply strip battery cable, slide compression nut onto cable and screw nut into terminal. Photo courtesy of Chrysler Corp.

Left: Compression terminals for top and side post batteries. Right: Quick disconnect terminals use stud-type connection with wing nut for quick removal and installation of cables. Photo courtesy of Chrysler Corp.

chassis. The lug terminal is flattened and has a hole for an attaching bolt. In fact, it looks much like a large ring terminal.

Generally, battery-cable terminals are attached to the cable by crimping. This requires a special crimping tool. Small battery-terminal crimping tools are clamped in a vise or struck with a hammer to operate the crimping jaws. Larger types work much like the crimping tools used on small terminals, except they're much bigger.

Some universal terminals are installed to the cable by a clamping screw or set of screws. These are best reserved for emergencies—they're far too crude for a professional installation.

Another type of terminal attaches to a cable by means of a *crimping nut*. A cable is slid into the terminal and the crimping nut is tightened. It works on a principle similar to that of a ferrule fitting on a copper water pipe. It provides a quality, finished appearance.

MISCELLANEOUS HARNESS SUPPLIES

This section covers all the miscellaneous components you'll need to make a wiring harness—except the wire, which was discussed in an earlier chapter. These include grommets, insulators, harness wraps, ties and clamps.

Grommets—When making a hole through a body panel for routing wires, use a grommet to cover the ragged edges. Although this is mandatory for holes made in sheet metal, it is often over-looked for those in fiberglass. This is a mistake for two reasons. First, fiberglass strands can eventually cut through the wire insulation. If this happens to adjacent wires, there's potential for a short circuit. Second, a correctly fitted grommet provides protection against the intrusion of water, fumes, air and dirt. If the fit is not close enough, seal the openings with silicone seal to ensure a weather-tight fit.

Grommets are usually made of rubber or a rubberlike material. There's a groove in the outside diameter that snaps over the edge of the hole in the panel. The wires pass through a hole in the center of the grommet.

In selecting a grommet, consider several dimensions: First, the diameter of the hole in the grommet must be just large enough to contain the wires. The hole in the panel must be just large enough to accommodate the diameter of the *bottom* of the groove. Finally, the groove must be just wide enough to fit the thickness of the material the hole is cut in. With fiberglass, this last dimension is difficult to match because most grommets are made for sheet metal, which is thinner than most fiberglass.

If you can't find a grommet to fit your needs, wrap enough layers of electrical tape around the wires to wedge them into the hole. Allow enough slack on both sides of the panel so the wires are not strained against the edge of the hole.

Insulators—An insulator is any material that prevents shorts between adjacent wires or between wires and grounds. Obviously, automotive wire is manufactured with an insulative covering—usually PVC plastic—so it doesn't need additional insulation. But wire connections may.

Shrink wrap, also called *heat-shrink tubing,* is commonly used to insulate wire connections. It is manufactured in a variety of lengths and diameters, and comes in several thicknesses. The least expensive type is a thin-wall plastic tube that shrinks to about half its original diameter when heated.

High-quality, thick-wall shrink wrap is preferable because it is less likely to split when heated enough to yield maximum shrinkage. A thick-wall shrink

Grommets come in numerous sizes to accommodate a single wire up to a large bundle.

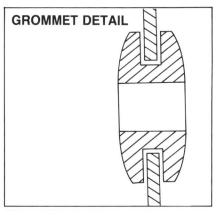

GROMMET DETAIL

When drilling hole in panel for grommet, make sure panel reaches bottom of grommet groove. Choose grommet with a groove width that matches panel thickness. To minimize wind or water leaks, wire bundle should fit grommet opening snugly.

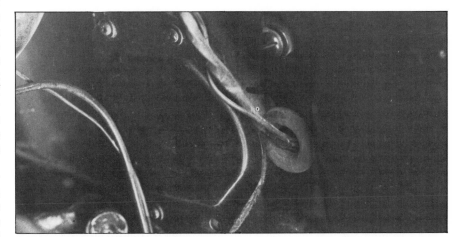

Use grommet to protect wires that pass through firewall or panels in bodywork. Photo by Jim Barrett.

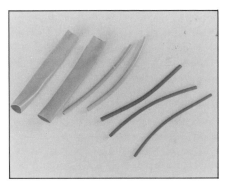

Shrink wrap tubing comes in many sizes and colors. Use to insulate and reinforce junctions of wires. See instructions on pages 116-118. Photo by Ron Sessions.

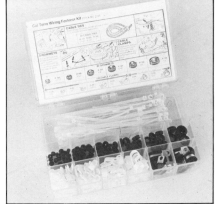

Wire fastening kit from Cal Term includes assortment of nylon ties, grommets, nylon clamps and rubber-insulated metal clamps. Photo by Ron Sessions.

wrap also provides extra support at terminal ends to prevent wire fatigue at the connection point. The most expensive type contains a hot-melt adhesive that runs into the connection to provide additional protection against moisture penetration. Alpha FT-105 is one brand of high-quality, thick-wall shrink wrap—there are others. Check the supplier's list on pages 152-153. For information on installing shrink wrap, see pages 116 and 118.

Electrical tape is another form of insulation. I mentioned it not because it is a *good method*, but simply to give me the opportunity to beg you to resist using it. Moreover, it looks terrible, it is susceptible to heat, and the adhesive may deteriorate in time, causing the insulation to drop off unexpectedly. Electrical tape is not recommended for harness wraps, either.

Harness Wraps—To bundle wires into harnesses, you can either use nylon ties, described below, or one of the harness wraps described here. Each method has advantages and disadvantages. Using ties is perhaps the easiest way to bundle a harness, so this is the bundling method described in the chapters on making and installing a harness. On the other hand, a correctly installed harness wrap can give the harness a neater appearance than ties. Wraps also provide additional protection from heat, grease, dirt and abrasion.

Split loom is a flexible, corrugated conduit made of black polyethylene. The conduit is split lengthwise for easy insertion of wires. After the wires are inserted, the seam springs back to a closed position and remains tightly closed.

Loom tape is a non-adhesive vinyl tape used specifically for wrapping harnesses. Because it is non-adhesive—it adheres to itself and nothing else—a piece of shrink wrap is used at each end of the wrap to keep the tape from unraveling. For more information on installing loom tape, see page 141. As mentioned, electrical tape or other adhesive-type tapes make poor harness wraps.

Ties—Nylon ties are narrow strips of plastic with a row of teeth molded into their entire length. They are available in many widths and lengths. At one end, a clamping block grips the teeth of the tie when the tie is pulled through. They are easily installed without tools. However, they are one-way; that is, they can't be loosened easily, so you must cut them to remove them.

Nylon ties are used to bundle wires into harnesses. Because the wires are not under much strain after the harness is installed, don't go overboard with size. Because width and length increase together, I found that 6-inch ties are long enough, though a little wider than need be. But, narrower ones are often inconveniently short.

Tie mounts allow you to use ties to attach the harness to flat surfaces. The mount is a small plastic block with a slot through it for the nylon tie to pass through. The back of the mount is self-adhesive. A stronger mount includes a hole for a #8 screw to attach the mount.

Harness Clamps—There are several types of clamps and clips to attach the harness to the car's frame or body.

Wiring straps are plastic straps with prongs or legs on one end that push into a predrilled hole. The other end has a barb. After wrapping the strap around the wires, install the barb into the hole provided. The barb expands the prongs behind the panel, locking the strap into place. It's possible to pry one of these straps loose without destroying it, but they're not designed for frequent disassembly. Wiring straps are available in a variety of configurations that all work essentially as described here.

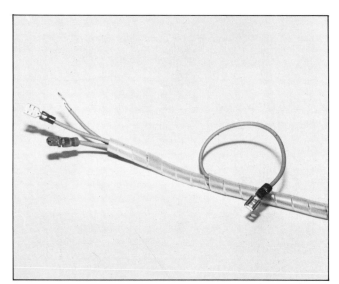

Spiral wrap is a good alternative to corrugated split loom. It permits branches to be placed anywhere along length of harness. One drawback is that it has to be "screwed" onto wire bundle. This is time consuming and makes it inconvenient to add or remove a wire after the harness is finished. Photo courtesy of Chrysler Corp.

Factory harnesses are usually covered with corrugated split loom. It is also available in the aftermarket in a wide variety of sizes.

Rubber-insulated metal clamps and *nylon cable clamps* are excellent for attaching wires and harnesses to most surfaces. Each type requires a single hole in the panel or frame rail for attachment. The clamp can be installed with a nut and bolt or with a rivet. Either method is fine, but a Pop rivet is less time-consuming to install and costs less.

One drawback of these clamps is that they cannot be adjusted for tightness except by replacing them with a different size. Another drawback is that the holes in the larger-size clamps are too large for rivet attachment—the rivet head will pull through. In this case, you'll need a backup washer under the rivet head.

Push-on frame clips are good for attaching harnesses to the edges of thick panels or chassis rails. These spring-steel clamps can be clipped on without using tools, and they're fairly easy to remove or adjust. The number of wires they'll carry is limited, so they are best used for fairly small bundles, or for single or double wires running to a remote location on the car. They are best for adding an accessory circuit to a car that's already wired.

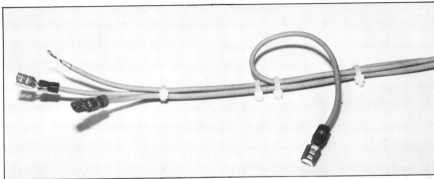

This example demonstrates correct use of nylon ties at a branch point. To keep bundle neat, install tie on main harness immediately on either side of branch. Space ties about 6 inches apart along harness.

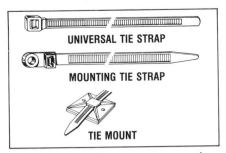

Nylon ties are available in numerous sizes and several colors. Length and width increase together as you progress up through the sizes. Ties are easy to install and remove with wire cutters, but cannot be reused. Everyone has a preferred harness wrap. Mine is nylon ties. They can also be used with nylon tie mounts to attach harness to car.

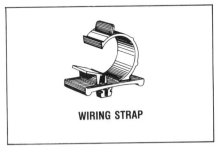

Plastic wiring strap has prongs that push into predrilled hole. Barb at other end pushes into slot, spreading prongs.

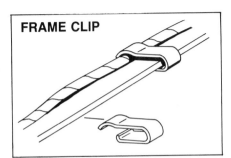

Push-on frame clip attaches wire or harness to edge of chassis flange. No tools are needed.

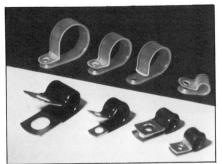

Nylon clamps (top) provide good isolation for harness as well as protecting wire insulation. Rubber-insulated metal clamps (bottom) are superb for tight retention of harnesses. Both kinds are installed to flat surface with rivet or screw.

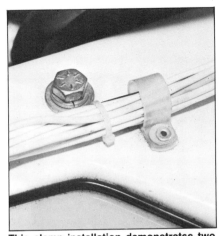

This clamp installation demonstrates two limitations. First, clamp is slightly too large for a good fit—the next smaller size was too small. Second, notice that I had to use a washer behind the rivet. Hole in clamp was too large and rivet would have pulled through.

Locating & Installing Components

Designer of my component car determined where headlamps would go, but I still had to make the final adjustments.

When a Detroit car designer works on a new idea, he first tries it on large, full-size drawings. He checks for interferences, operating space and other important factors. This is time-consuming, but the drawing board is the cheapest place to make mistakes and find solutions. Next comes the hand-built prototype car, costing hundreds of thousands of dollars. Here, the designer's concepts are transformed into a full-size mock-up, where any oversights are caught and corrected. After revisions and more tryouts, the new car design is put into production. If a mistake makes it this far, it will be repeated thousands of times!

Your project car is similar to that first prototype. You probably won't have blueprints to work from, so your concepts become permanent during the tryout stage. So, you should work carefully and deliberately to avoid oversights that you'll have to live with for thousands of miles.

CHOOSING COMPONENT LOCATION

Due to manufacturer's design or legal requirements, some electrical components must be installed in a specific location. On kit cars, for instance, body design and engine requirements dictate the basic location of many components. These are called *fixed-location components*.

Depending on the project car you're building or modifying, other components can be installed practically anywhere you want. These are called *optional-location components*. A few components fall between these two categories, as discussed later.

Fixed-Location Components—As mentioned, where you install fixed-location components is usually determined by legal requirements, vehicle design, or both. Headlamps are one example. There are minimum height requirements in every state. If you're building a kit car, the kit design further

dictates headlamp location—in the grille, fenders, separate buckets or whatever. So, you often don't have many choices—or any—in determining headlamp location. Other components, such as taillamps, also fit this category.

Optional-Location Components— These are components whose location is not dictated by law or car design. The fuse panel is one example. Within reason, you can install it in any convenient location—inside the glove box, behind the dash or under the seat, for example.

Fixed or Optional?—There are some *gray areas*—components that may have a designed location but can be installed elsewhere. The battery is one. For instance, a kit car may include a rectangular battery well in the engine compartment. But you're free to put the battery in another location if you want, provided there's good ventilation and enough room.

With all components—fixed, optional or in-between—you still must follow the

Size, number and location of taillamps is often dictated by local or state laws. Be sure location you've chosen will conform.

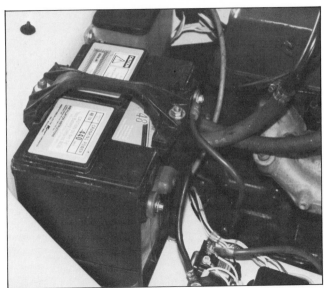

This battery location was dictated by two factors. First, the fiberglass body was molded with battery shelf in place. Second, this location places battery within about 18 inches of starter motor, minimizing cable length.

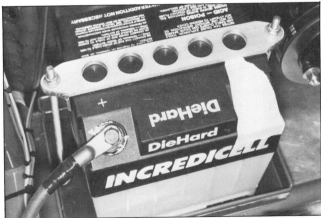

In his street legal sprint car, Ron Fournier located battery behind seat in tail of car, for improved weight distribution. This also placed battery within a few feet of rear-mounted alternator.

Another example of weight distribution: battery for front-engine Trans-Am Mustang is mounted at rear, along with fuel cell and dry-sump oil reservoir. The sacrifice: long cable run from battery to engine. Photo by Tom Monroe.

basic principles of observation and test-fitting before permanently installing them. This is true even if a kit-car manufacturer has designated specific locations for components. That being said, all components discussed from now on will be lumped into the general category *components,* with none of the above distinctions being made between types again.

Mechanical Factors—Like the fixed-location components just discussed, components influenced by mechanical factors are located by considerations beyond our control. *Function* and *environment* dictate the basic location of these components.

Obviously, the starter motor, distributor, and other such engine-mounted components cannot be moved from their designed location. But consider the other related mechanical components. For instance, if the starter motor requires a separate starter relay, the relay must be installed close to the motor. You may pick the exact spot on the inner fender or firewall for the relay, but physical requirements determine its general location.

The same principle applies to the ignition coil. It may be mounted to the cylinder head, engine block or inner fender at your discretion, but it still must be in close proximity to the distributor.

It doesn't take much thought to realize that wiring-harness location is in-

LOCATING AUXILIARY LAMPS

Just as headlamp location is dictated by state regulations, so are the number and location of auxiliary lamps—that is, unless you're building a car for strictly off-road use. But if you plan to drive the car on a public street, road or highway, the lamps must conform to the laws of your state. And if you drive out of state, they're subject to the laws of states in which you travel.

As the accompanying chart shows, mounting heights and maximum allowable number of lamps vary from state to state. Most states also have additional regulations that aren't covered in the chart. For instance, roof- or rollbar-mounted lamps are illegal in almost all states, but are tolerated in some if the lamps are covered and disconnected—usually by pulling the fuse—when on public roads. So, the best advice is to check with your local motor vehicle department or highway patrol office before installing auxiliary lamps.

Here are some mounting tips:
• Mount fog lamps and short-range driving lamps as low as practical, preferrably below headlamp level. This minimizes glare and reflection in fog and dust, and better illuminates the road directly in front of you.
• Mount long-range driving lamps as high as practical, preferrably above headlamp level. But remember that roof-mounted lamps are illegal in most states. They also have a tendency to reflect off the car's hood.
• Mount lamps where they will be protected. This is more critical on race cars, especially off-roaders, than for street vehicles. But lamps mounted below the front bumper or protruding past it are asking to get smashed.

Most states have laws governing mounting heights of auxiliary lamps. Check these before installing yours. This fellow may get cited! Photo by Jim Barrett.

LIGHT-MOUNTING RECOMMENDATIONS

This chart indicates the mounting specifications preferred by the AAMVA (American Association of Motor Vehicle Administrators) and should be consulted prior to mounting auxiliary lights. Chart courtesy of KC HiLites Inc.

State	SPOT LAMPS Max. No.	FOG LAMPS Max. No.	FOG LAMPS Mtg. Hgt. (in.)	AUX. PASSING LAMPS Max. No.	AUX. PASSING LAMPS Mtg. Hgt. (in.)	AUX. DRIVING LIGHTS Max. No.	AUX. DRIVING LIGHTS Mtg. Hgt. (in.)
Alabama	1	2	12-30	1	24-42	1	16-42
Alaska	2	2	12-30	2	24-42	2	16-42
Arizona	1	2	12-30	2	24-42	2	16-42
Arkansas	2	2	12-30	2	24-42	2	16-42
California	2(1)	2	12-30	2	24-42	2	16-42
Colorado	2	2	12-30	2	20-42	2	16-42
Connecticut	2	2	12-30	2	24-42	2	16-42
Delaware	2	2	12-30	1	24-42	1	16-42
D.C.	2	2	12-30	2	24-42	2	16-42
Florida	1	2	12-30	2	24-42	3	12-42
Georgia	1	2	12-30	1	24-42	1	16-42
Hawaii	—	—	—	—	—	3	12-42
Idaho	2	2	12-30	2	24-42	2	16-42
Illnois	1	—	—	—	—	3	12-42
Indiana	2	2	12-30	1	24-42	1	16-42
Iowa	1	—	—	—	—	3	12-42
Kansas	2	2	12-30	2	24-42	2	16-42
Kentucky	—	—	—	—	—	—	—
Louisiana	Prohibited	2	12-30	2	24-42	2	16-42
Maine	1	2(2)	—	2(2)	—	2(2)	—
Maryland	1	2	12-30	—	—	2	16-42
Massachusetts	1	—	—	—	—	—	—
Michigan	2(5)	2	12-30	—	—	2	24 Min.
Minnesota	2	2	12-30	2	24-42	2	16-42
Mississippi	1	—	—	—	—	2	24 Min.
Missouri	1	—	—	—	—	3	12-42
Montana	2	2	12-30	2	24-42	2	16-42
Nebraska	1	—	12-42	—	—	2	24 Min.
Nevada	2	—	—	—	—	2	16-42
New Hampshire	2	—	—	—	—	3	12-42
New Jersey	1	—	—	—	—	2	12-42
New Mexico	2	2	12-30	1	24-42	1	16-42
New York	—	—	(6)	—	—	—	—
North Carolina	2	—	—	—	—	2	—
North Dakota	2	2	12-30	2	24-42	2	16-42
Ohio	1	2	12-30	2	24-42	2	16-42
Oklahoma	2(3)	2	(4a)	—	—	—	—
Oregon	1	—	—	—	—	3	12-42
Pennsylvania	1	—	—	—	—	(4a)	12-42
Rhode Island	2	—	—	—	—	2	(4)
South Carolina	1	2	12-30	1	24-42	1	16-42
South Dakota	1	—	—	—	—	3	12-42
Tennessee	—	—	—	—	—	—	—
Texas	2	2	12-30	2	24-42	2	16-42
Utah	2	2	12-30	—	—	—	—
Vermont	—	—	—	—	—	—	—
Virginia	2	2	—	1	—	1	—
Washington	2	2	12-30	2	24-42	2	16-42
West Virginia	1	2	12-30	1	24-42	1	16-42
Wisconsin	2(3)	2	12 Min. (4ab)	—	—	—	12 Min. (4ab)
Wyoming	2	2	12-30	1	24-42	1	16-42

(1) Cannot exceed 32Kcp or 30 watts
(2) Total of two fog or aux. lights permitted
(3) Must be mounted at height between 30- and 72-in.
(4a) Below headlight centers

(4b) Not to exceed 75Kcp
(5) Emit white or amber light only
(6) Affix amber lights below headlights and avoid glare or dazzle

fluenced by mechanical factors. Harnesses should take the shortest practical route to the components they serve. Harness location is further dictated by structural members to which the harness must be attached. Harnesses cannot be routed across open spaces, and they must be protected from engine heat and moving parts.

So, the important considerations of mechanical factors are *component function* and *component environment*. These considerations are further discussed under the heading Clearances, below.

Although some preceding remarks about mechanical factors may seem rather obvious, they were made for a good reason. They'll help enable you to distinguish mechanical factors, which you cannot control, from *human factors,* which you can and must control.

Human Factors—The science of *human engineering* deals with how various human factors—anatomy, physiology, psychology and so on—relate to the design and positioning of machines, instruments and controls. Its aim is to design products that will serve the user in the intended way. Here, the term applies to how you relate to the controls and displays in your car.

Human engineering is both complicated and given to a lot of specialized buzzwords. However, some basic principles can be used to help you determine the locations of electrical components in the passenger compartment of your project car.

There are two basic ideas to consider when you're designing for human factors. The first is to be sure the design permits maximum effectiveness and efficiency when you use a control or look at a display. The second is related to your safety while operating the device or looking at the display. Can you do so without causing something undesirable to happen?

For instance, if the headlamp switch and windshield-wiper switch are next to each other, have the same shape and operate the same way, it may be possible to turn the headlamps off instead of turning the wipers on. This could be disastrous on a dark, rainy night! Such a design violates the rules of good human engineering, or *ergonomics.*

This fuse panel is located under instrument panel next to driver's right knee, where there was plenty of room for panel and associated wiring. It is easily accessible, close to electrical distribution panel (terminal strips with 92 connections), and protected from accidental contact by moving parts.

Ideal horn location is near the front, protected from direct road splash and away from suspension, steering and engine-fan components. Mine is mounted at bottom of engine compartment, bracketed to chassis rail near front crossmember.

So, consider human factors early in the design process, and do it in a systematic way. These factors also apply to selecting gauges, indicators and controls. Controls are discussed on pages 28-34, gauges and indicators on pages 38-48.

CLEARANCES

Note: This section is concerned with clearances for component installation only. It does not cover wiring, except as it relates to clearances for connection to components. Clearance for wiring-harness installation is discussed on page

Allow enough room to route and connect wires to components. These wires attach to six terminals of DPDT wiper switch on my car. This photo makes it look worse than it actually is, but the wires still constitute a fairly large bundle to deal with in close quarters.

Windshield-wiper motor is a fairly large, heavy device to position. It has quite a few wires to service it—don't forget to provide ground wire if motor is mounted to fiberglass. Be sure to allow enough room for operation of attached mechanical arm and linkage—it is powerful and can actually break things that get in the way.

102. This chapter deals with the functional location of the components the wires will eventually attach to. Wiring requirements are covered in the following three chapters.

A poorly fitted component may look bad, function poorly or be inconvenient to operate. It may also be difficult or impossible to service, or get in the way of servicing a neighboring component. The obvious cause of such problems is *poor planning.*

Determining clearance for a component consists of asking yourself a

Alternator location is usually dictated by engine manufacturer. OEM alternator brackets are heavy castings or stampings. Here, I used top alternator bracket for mounting ignition coil to keep it close to distributor.

Before I saw this, I'd never thought of mounting an alternator anywhere but on the engine. Race-car fabricator Ron Fournier wanted to keep the engine compartment of his sprint-car-type street rod authentic—sprint cars don't carry an alternator. He fabricated mount and drive to permit rear of drive shaft to run alternator. One drawback is that alternator puts out only when car is moving.

This starter solenoid is ideally located—within inches of battery, and within about 14 inches of starter motor, minimizing cable lengths.

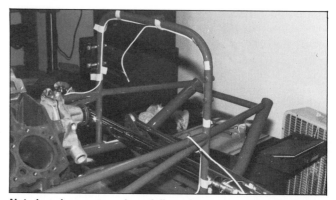

Note how harness mock-up follows hoop shape of radiator support. It would have been shorter to run harness straight across, but it would have been vulnerable to damage and have looked terrible.

Ford brake switch is designed to be located directly on brake pedal arm. I had to add a wiring clamp to keep wires from getting pinched by up-stop of clutch arm next to brake.

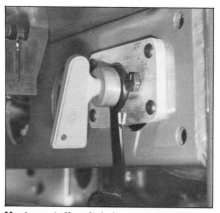

Master cutoff switch is used to kill car's entire electrical system. Large handle on switch is removable for security. Because this car has computerized fuel injection system, we provided an always-live circuit, bypassing this switch, to keep computer's long-term memory alive.

Location of electric fuel pump is well planned—below fuel-tank level for good gravity feed to pump, close to fuel tank to help prevent vapor lock on hot days, and well protected within car's framework.

series of questions. How well you answer them, or whether you answer them, will largely determine how well the components fit into their designated spaces and how well they function. There are four clearance considerations in planning component locations: *clearance for installation, clearance for associated wiring, clearance for operation* and *clearance for service.*

In a practical sense, all four clearances are interrelated and overlapping. So, it's necessary to define them only to make you aware of the great number of factors that can cause a component not to fit.

Also associated with these clearance factors are the ideas of *careful examination, careful planning* and *careful test-fitting.* These will be covered after I explain the four clearance factors.

Clearance for Installation—This is simply a matter of space. Is there enough room for the component where you want it to go? Are there any structural members in the way? Can doors, hood and other *movable* body parts operate through their full range without interfering with the component you're installing? Can the component operate without interference from adjacent components? See Clearance for Operation, at right. Is there a way to get the component into the position you've selected? Does the component have to be installed before or after

another one to avoid interference during installation?

Clearance for Wiring—Once the component is installed, can you get the associated wiring to it? Is there enough room for the wire connectors? Is there enough room for your hands and tools to install the connectors? Wires and connectors are bulkier than they look when space is cramped. Is there enough room for the wiring harness? The harness bundle can become bulky and difficult to bend when space is limited.

Clearance for Operation—If the component is a switch or knob, can you reach it from the normal driving position? Is there enough room for your hands and fingers to operate it without interference from adjacent components? Can it be operated when you're wearing gloves or mittens? Does the switch handle contact an adjacent component in any of its positions?

If the component has moving parts, is there room for these parts to operate in all positions? A good example is the linkage on the back of the windshield-wiper motor. If there are several moving parts close to each other, all of them should have room to operate freely, regardless of position.

Clearance for Service—If the component eventually fails—it eventually will, you know, and at probably the least

convenient time—will you be able to replace it? Is there room around the component to make adjustments? Is there room to unplug the wire connectors for test purposes?

GUIDELINES FOR SPECIFIC COMPONENTS

Many of the components listed in this section will be optional-location components in your car—perhaps all of them. If you're installing optional-location components that aren't mentioned here, follow the general guidelines just discussed—and use common sense.

Battery—As mentioned earlier, the battery may have a design-limited location. If not, locate after considering the following:

● Can it be securely fastened down?

● Will it be easy to get to for checking and adding water—unless it's the newer, sealed-for-life type?

● Is it close to the starter motor? This keeps long, heavy battery leads to a minimum and reduces unnecessary voltage drop.

Gauges—On kit cars, gauge location may be dictated by kit design—particularly in a replicar, where authenticity may be important. On production cars, the location of add-on gauges may or may not be dictated by the

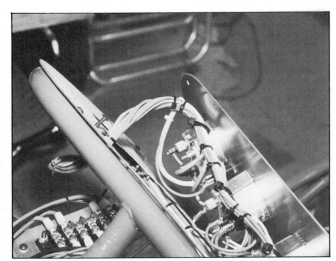

Instrument panel tips out for easy service. We provided loop in gauge feed wires to accommodate this feature. As you can see, wiring was done before car's outer skin was installed, making it a snap.

Race car's angled instrument panel provides optimum gauge visibility and switch access from driver's seat. Dzus 1/4-turn fasteners allow quick panel removal for pit repairs. Tach is positioned so red line is straight up. Photo by Tom Monroe.

instrument-panel design, depending on the car and the gauge being installed. Even though gauge location may be dictated, refer to the following list to be on the safe side. If you have a choice, be absolutely sure to consider these:

● Can the speedometer cable be routed without kinking?

● Can you see the gauge faces clearly from the driver's seat? This is why it's important to have the seats and steering wheel installed, at least temporarily, before determining the location of any of the gauges.

● Are the most important gauges in the most important places? Only you can decide if the tachometer is more important than the speedometer, or the oil-pressure gauge more important than the coolant-temperature gauge.

● Are there provisions for lighting the gauges? If gauges don't come with built-in lamp sockets, you must determine the appropriate locations for illumination lamps.

SWITCH SPACING

Preferred distance in inches; minimum distance in parentheses.

	Knob	Push Button	Toggle Switch
Random Operation	2 in. (1in.)	2 in (1/2in.)	2 in. (3/4 in.)
Sequential Operation	—	1 in. (1/4 in.)	1 in. (1/2 in.)

● Is there room for the gauge wiring behind the instrument panel? Is there sufficient access behind the panel for hands and tools to install the gauges and attach the wires? If not, it might be easier to install all gauges in a removable panel with the gauge leads already attached. You can always run these pre-attached wires to a junction block for continuation of the circuits.

Indicator & Warning Lamps—These deserve many of the same basic considerations that you give to gauges: Make sure you can see them from the driver's seat and that they aren't hidden behind the steering wheel. Is the layout of the lamps logical? For instance, the left-turn indicator should be located to the left of the cluster and the right turn indicator to the right. If you're using lamps in conjunction with gauges—the oil pressure gauge and warning lamp, for instance—the related components should be close to each other for quick discrimination.

These switches are widely spaced for minimum of confusion, even though both pairs are alike. Toggles operate circuits that are normally ON when the car is running. Flip-up safety shields prevent accidental operation of toggles. Push buttons are *momentary-on* type.

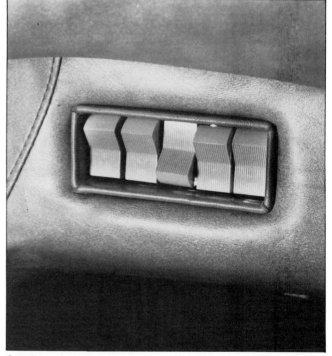

Spacing of rocker switches in my car is not ergonomically sound—they're much too close together, are all alike in appearance and touch, and are not labeled. I chose these switches primarily for appearance—not a good ergonomic reason! The reason they're not labeled is to maximize confusion to potential tamperers—it's an open roadster with little intrusion protection.

Switch safety shield flips up to permit turning switch on. This two-step procedure minimizes accidental operation, though in an emergency, switches can be turned off instantly just by pushing on shields.

Off-road racers prefer widely spaced switches—less chance of driver hitting wrong switch while bouncing over rough terrain. Photo by Tom Monroe.

Switches—The locations of various switches on the instrument panel may be dictated by car design, components the switches serve, or their logical placement in relation to other switches and components on the panel. Logical placement as an important consideration in locating switches. Don't forget this and end up with switches placed in random locations all over the panel. It will not only be unattractive, but will be inconvenient when you operate them. Here are some considerations:

After general locations are determined for switches, you must arrange them in specific locations. Often, it's helpful to position controls for related systems in the same general area. Two things to give careful thought to are the layout of controls relative to each other, and spacing between them.

If there is a sequence of operations that is commonly used for a set of controls, lay them out to make that sequence convenient to follow. A good example is the mode, temperature and fan controls for a heater or air-conditioning system. Another is the windshield-wiper and washer controls if these are controlled by separate switches.

The spacing of controls is just as important as location. It's desirable to space controls as close together as practicable, simply to minimize unnecessary reaching—especially in the case of the related controls mentioned above. But

there's a *minimum distance* to observe so you don't accidentally operate a control by mistaking it for another one. This is especially the case in off-road racing and rally cars where the going is extremely rough. It's not unreasonable to double the recommended distances to avoid accidental activation of a switch.

The chart on the facing page gives typical distances between controls. Measurements are taken edge to edge. Preferred distance is given in inches, minimum distance in parentheses. Here are some other considerations for locating switches and controls:

● Can the switch be reached comfortably from the normal driving position?
● Is there sufficient clearance on all sides of the switch?
● Is there sufficient room behind the panel for attaching the wire connectors?
● Will the connectors come into accidental contact with metal components, which could lead to short circuits, fires and other not-fun occurrences?
● Is there a need or provision for illuminating the switch? If so, is there enough space for the lamp and associated wiring?
● If the instrument panel is fiberglass, will the switches you've chosen mount to the panel? Remember, fiberglass is usually much thicker than sheet metal usually found in the panels of factory installations.

INSTALLING FIXED-LOCATION COMPONENTS

The time to answer the questions raised so far is *during the test fitting*, not after you've cut openings or drilled holes.

Test Fitting—If you're building a kit car, you might assume that fixed-location components don't need to be test fitted. After all, there's only one fixed place for certain components, so you may as well go ahead and cut or drill. The kit manufacturer put dimples in the body just where taillamps are supposed to go, didn't he? But stop and think about it for a minute.

What material are most component-car bodies made of? Fiberglass. And how is the thickness of fiberglass controlled? Often, it's not. This is an inherent characteristic of the material and the process, not necessarily an indictment of the body builder. So, don't take anything for granted.

Many fixed-location electrical components depend on major parts of the car being assembled before you install electrical components. So, I've assumed your car is essentially assembled, with body parts, engine, suspension and major interior components installed, at least temporarily.

Decide which fixed-location component to install first. If the component is to be mounted on the body or in the passenger compartment, cover the area

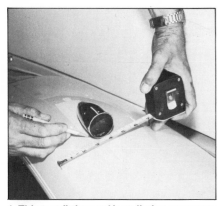

1. This test fitting and installation sequence for fender-mounted turn-signal lamp shows the basic routine. Modify it to fit your particular situation. Begin by measuring. If component is duplicated on opposite side of car, measure both. Tape on fender protects finish and provides marking area.

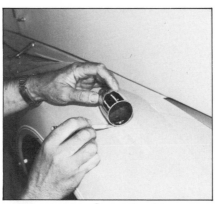

2. Trace position of turn-signal lamp onto tape. Be sure to check position visually as well as double-checking your measurements.

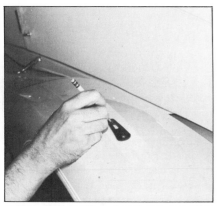

3. Here, I've positioned gasket over outline of lamp base. Again, accuracy counts. Gasket must be perfectly centered. Trace lamp-mounting and wire access holes, using gasket as a template.

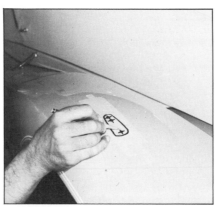

4. After outline of gasket and holes are marked, mark centerlines for holes to be drilled.

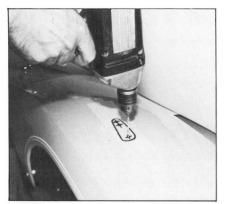

5. Drill holes. In fiberglass you may want to predrill with a small bit run a slow speed. Then follow with final-size bit at moderate speed. Be careful not to permit rotating drill chuck to contact tape or tear through it.

6. When I mounted my lamps, I found that the screws furnished with them were so long that they actually touched the lamp bulb. Fearing eventual lamp failure, I filed screws shorter. Don't be surprised if things don't always fit exactly the way they're supposed to.

with masking tape to protect the finish. The tape also provides a convenient place to draw the centerlines of holes and outlines of openings. Then, have an assistant hold the component in place while you step back to view it from every angle possible.

If you don't have an assistant—you'll probably have several, because project cars often draw kibitzers like flies—temporarily brace or tape the component in place. I was thinking specifically of free-standing headlamps when I wrote this, but the same applies to most other components.

If you are building a kit car, check to see that the component is positioned according to the kit manufacturer's instructions. Does the component look right where it is? Will you need to make any adjustments or allowances for production tolerances? If the component looks good in its position, check clearances of moving parts and other clearance considerations.

Installation—When you're sure the component fits the location you've chosen for it, the next step is to measure and mark on the masking tape the location for the hole or holes. But, before you drill or cut any holes, *double check your measurements*. The first and last rule of installing components is: *Measure twice, cut once.*

Matched components on a car, such as headlamps and taillamps, are mounted symmetrically. That is, they're mounted in seemingly identical spots on the opposite sides of the car. Again, because of the nature of the fiberglass body material and the various levels of workmanship, you may or may not find that symmetry actually exists. So, to ensure best appearance, measure for each component separately, but test-fit symmetrical components together.

Installing components correctly involves pre-planning, checking, rechecking and finally doing it. At this point, follow the installation instructions provided with the component or the car kit. If no instructions came with either, refer to the appropriate shop manual. If you're reasonably careful, the end result will be a joy to behold.

INSTALLING OPTIONAL-LOCATION COMPONENTS

When all fixed-location components are installed, determine the exact locations for the optional-location components and install them. Don't install these until you've thought everything through and checked each location.

Test-Fitting—For optional-location components such as the battery or bracket-mounted gauges, follow the test-fitting procedures given for fixed-location components. For hole-mounted gauges, indicator lamps and switches, make a full-size *mock-up* of each component. This is especially helpful in determining exact spacing of components that require large holes. For indicator lamps, remove the lenses or the bezels from the lamp assemblies, if possible, and temporarily tape them to the instrument panel. For switches, you may be able to remove the bezel or mounting nut. Otherwise, make a cardboard mock-up.

When you make your gauge mock-ups you don't have to create works of art, but they should be fairly representative of the actual gauge appearance. You can use an inexpensive pencil compass for drawing the circles on a piece of thin cardboard. A strip of red tape can be used to represent the gauge needle. And be sure to include the outside diameter of the *gauge bezel* in your mock-up. If you don't, you may position the gauges too close together.

Tape the mock-ups into the locations you've selected for them on the instrument panel. Moving a taped-on mock-up at this point is much easier than moving a 2-inch hole later! Sit in the car seat and carefully stare at each gauge and indicator in the spot you've chosen for it. Reach out and pretend to operate each switch. Repeat this drill until you're actually tired of doing it. Move the components around until you think you're satisfied with their locations.

When you've finally decided on the exact location of each gauge, indicator and switch, walk away from the car for awhile. When you come back and see it with fresh eyes, you may notice something that you overlooked. And, if there are going to be several drivers, have each sit in the driver's seat for a thorough critique. This is your last chance, so take your time. Take one last measurement

Mock-up for race-car instrument panel is done with steering wheel and seat temporarily in place. Whether you're designing the panel or installing gauges to an existing panel, you can either draw gauge mock-ups, or mount bezels or actual gauges on cardboard, as shown here.

behind the instrument panel for each component, just for insurance.

Eventually, you'll be satisfied with your chosen locations, and the moment of truth has arrived. Carefully punch a hole through the exact center of each mock-up. Through this hole, use a center punch or small drill to indent the surface of the instrument panel.

Drilling Holes—Remove the mock-ups, one at a time, and write the size of the necessary hole next to each center-punch mark. The hole for the component will be *smaller* than the outside diameter of the bezel, so measure and mark the holes accordingly. Finally, use the correct-size drill bit or hole saw to cut each hole, referring to the pencilled dimensions.

For rectangular holes, drill a small hole at each corner of the cutout, then cut between the holes with a sabre saw. Cover the underside of the saw's baseplate with several layers of masking tape to keep the saw from marring the panel surface. Finally, use a small file to square each corner of the cutout.

For steel panels, use a hole saw or sabre saw with a fine-tooth metal-cutting blade. I've found that a fine-tooth wood-

State laws require turn signals; authenticity required not having handle show on steering column. So, we mounted this turn-signal switch under dash so only tip of handle is visible from above. This is the kind of aftermarket switch designed to install in older cars and trucks that did not originally have turn signals.

cutting blade works well for fiberglass. Regardless, be sure the blade is sharp—fiberglass tends to dull cutting edges quickly.

If you're working with a fiberglass panel on a kit car, make some test holes to check hole shape and size. Make test holes in a scrap of fiberglass removed from the car body. For instance, many kit cars are fabricated with reinforcing webs in door openings, which you'll be removing and discarding anyway.

I wanted to wire my car with five separate harnesses, joined by terminal blocks. There was enough room behind center console to mount the blocks, so I fabricated a mounting plate for both blocks and fuse panels.

Several dozen wires will pass through this opening for fuse panels. I split a piece of rubber hose and installed it on the raw edge to protect wires.

The three fuse panels neatly span the opening.

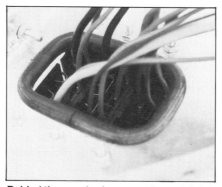

Behind the panel, wires pass through fabricated grommet.

Mounting Components—Aftermarket gauges, indicator lamps, switches and controls are most often designed to mount to a flat surface; this is because there's no way for the manufacturer to anticipate all the specialized places you may want to mount them. Specific mounting procedures vary for different types of gauges, so the best advice here is to follow the gauge-manufacturer's instructions. General instructions for mounting common switches are given here.

Rotary and push-button switches mount from behind the panel, with the actuator protruding through a round hole. A threaded stem surrounding the actuator accepts a thin hexagonal or knurled nut, and a lock washer. Often, the switch includes a printed *switchplate* that can be mounted under the nut, if desired. This plate indicates switch-actuator position. A tiny tab on the plate installs into a machined slot in the stem for correct orientation.

When mounting these switches, be careful when you tighten the nut. It's quite thin, so the wrench may slip off and damage the mounting surface if you're careless.

Most toggle switches are hole-mounted like a rotary switch. They also usually have a switchplate. But some toggle switches and most rocker switches are *front loaded* into a specially shaped rectangular hole. The required dimensions of the hole are specified by the switch manufacturer. To cut a rectangular hole, follow the instructions in the previous section on drilling holes.

Because most front-loaded switches are retained by built-in spring clips, panel thickness is important. If the panel is too thin, it won't be tightly clamped between the clip and the switch body. You may need to fabricate a shim for correct fit.

A too-thick condition is most likely to occur in a fiberglass panel, such as the instrument panel. In this case, I suggest cutting a hole in the instrument panel large enough to accommodate all the switches. Then fabricate a metal insert of a suitable thickness and install the front-loaded switches to it. Dune buggies, race cars and vintage-replica kit cars look good with engine-turned or brushed-metal inserts, so you can add to the instrument panel's appearance at the same time.

When mounting toggle switches and rocker switches, orient them so they operate vertically. That is, up for ON, down for OFF. Use a horizontal orientation only if it is consistent with operation of the device. For example, you may want to use a toggle switch to operate the turn signals—though not likely—with the movement of the switch corresponding to the sides of the car. Another purpose for horizontal orientation of these switches is to prevent accidental movement, especially if the horizontal switch is mixed among vertical ones.

This chapter covers the installation of only the most common components. Other components and accessories, such as radios and speakers, are installed much the same way as the components just described. By now, you should know the routine—measure twice, cut once. Also refer to the installation instructions that accompany the component.

After you have all components installed to your satisfaction—and to the satisfaction of the requirements of mechanical factors, human factors and clearance considerations—you're ready to design the circuits that will tie these components together.

Both push button (left) and toggle (right) are installed from behind panel and locked in place with two nuts, one on either side of panel. This permits adapting these switches to panels of various thicknesses.

Designing & Drawing Circuit Diagrams

Symbols are the language of electrical circuits. Just as with a spoken language, the better you understand the individual elements, the better you will be able to communicate.

The preceding chapter describes how to locate and install the various electrical components in your project car. This chapter shows you how to make the circuit diagrams that you'll use as a guide to wire components together.

A glance at the wiring diagrams in a typical shop manual should convince you that a car's electrical system is complex. Viewed as a whole, it is. However, you won't need to make an overall schematic drawing of your car's entire electrical system.

When I was employed to develop electrical-diagnosis material for some of Detroit's auto-company shop manuals, I made a point of *isolating components* into independent, functional circuits. That made it easy to understand them, work on them and trace current paths in them. This is the approach used in this chapter to draw circuit diagrams for your project car. Because you'll be drawing each circuit separately, there's never a time during the design process when you

need to think of the car's entire electrical system at once.

In addition to the greater ease of understanding, there are other advantages to the individual-circuit approach. Because you're going to install each wire, one at a time, drawing each circuit separately will help simplify constructing the harness. Later, if you need to diagnose a problem, you'll have the separate circuit diagrams for reference.

SYMBOLS

If you've ever read a road map, you know how important it is to understand the meaning of the special symbols used. The same thing is true of circuit drawings. So, let's get started right by establishing some basic symbols to use for your diagrams.

Many different symbol systems are used for electrical-circuit diagrams. But for the sake of clarity, one set of simple symbols will be used throughout this

book. They are shown in the following pages. The reason for using these symbols is that they portray what's *inside* the component, so you can visually trace a circuit's various paths through that component. This is particularly helpful when depicting relays and multiple-pole switches, for example.

In recent years, many car manufacturers are using symbols of this type in their shop manuals. If possible, get the appropriate shop manual for the car model you're using as the donor of your electrical components. Chances are, if the car is relatively new, the manual will use symbols much like the ones used here. If the symbols are not exactly the same, you still should be able to decipher them by studying the sample wiring diagrams shown in this chapter.

To simplify the drawings, most components are represented by rectangles or circles. The shape of the symbol is not necessarily representative of what the component actually looks like.

WIRES

It seems logical to begin a description of specific symbols with the basic component of all circuits—the wires. Regardless of wire gage or insulation color, all wires are drawn as solid black lines. The exception is the phantom wire described later.

It's customary to write the wire gage in numerals next to the wire and to indicate insulation color, including the tracer color, if applicable. For example, a 12-gage green wire with a white tracer would be designated 12 grn w/wht, or something similar. If you haven't yet chosen the appropriate wire sizes and colors for the components you're installing, see pages 23-24.

Wire Colors—There's probably no universal agreement regarding abbreviations used in diagrams for the various wire colors. The diagrams in this chapter do not include wire colors because the car you're working on probably won't be wired exactly like the ones used as examples here. However, you should indicate wire colors on your diagram, so use the following abbreviations:

Color	Abbreviation
Green	grn
White	wht
Red	red
Blue	blue
Brown	brn
Black	blk
Yellow	yel
Orange	orng
Purple	prp

As described on page 25, a *tracer* is a stripe of a contrasting color that runs the length of a wire. Tracers are intended to provide a greater variety of "colors" to choose from. Factory wiring diagrams identify tracers in different ways. Diagrams that are printed in color simply duplicate the wire's appearance. Diagrams printed in black-and-white use labels. For example, here are some common designations for a red wire with a white tracer:

- Red w/wht (red with white)
- Red w/t (red with tracer)
- R/W (red with white)
- R/wht (red with white)

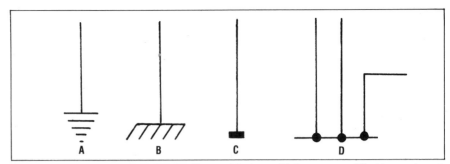

Most common ground symbol looks like upside down Christmas tree (A). Always draw it pointing down. In electronic work, symbol shaped like hair comb is used to indicate chassis (B). Many wiring diagrams for German cars use a straight line to indicate ground (C). Other diagrams represent car's chassis with straight line across bottom of diagram, with all the circuits grounded to it (D). Of all these, I find A to be most convenient and easy to understand.

Usually, the abbreviations used in factory diagrams are fairly obvious. Some diagrams also include a legend that identifies the abbreviations. In your diagrams, you can use any of the abbreviation methods above, as long as you keep them consistent in all your drawings.

Grounds—As mentioned in earlier chapters, cars are usually wired in a *single-wire* arrangement, where the chassis, engine block or metal body of the car serves as the return path to complete each circuit. That is, these parts act as *grounds*. In a diagram, the point where a wire or component is attached to the ground is represented by a symbol. Correct use of the ground symbol means you have far fewer wires to draw when designing circuits. This is because you won't be drawing wires to show return paths to the battery.

In most diagrams, the symbol used for ground looks somewhat like an upside-down Christmas tree, so that's what I use. It's made by drawing several horizontal lines, with the longest one on top and the shortest one at the bottom. The wire attached to the ground symbol comes in from the top, acting as the trunk of the upside-down tree. Ground symbols are shown in the drawing above.

You may find other representations for grounds in wiring diagrams for some imported cars. The diagrams used by German auto manufacturers often use a short, thick horizontal line. Other diagrams have a solid line running across the entire bottom of the diagram to

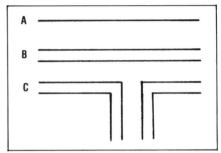

Wire is represented by a single line for each circuit (A). Where two or more circuits run parallel to each other, each one has its own line (B). Where wire changes direction on a diagram, draw a right angle (C); it's not necessary to get fancy and draw curves.

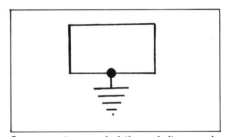

Component grounded through its case is said to be *self grounded*. The symbol for this includes a dot where ground symbol crosses outline of the component.

represent the chassis, and each grounded component is wired to it. To me, this is unnecessarily complicated and more difficult to draw, so don't use it.

Some components are *self-grounded,* meaning the body of the component itself acts as a conductor, completing the circuit to ground through its mounting bracket. On a self-grounded component, the ground is shown as a dot *on the component outline,* with a short ground symbol running from it. However, if a self-grounded component must be attached to a fiberglass or other non-metal body part, you must provide a *jumper wire* from the mounting bracket to a grounded metal part of the vehicle—usually the chassis or a designated large-gage ground wire running the length of the car.

FUSES & CIRCUIT BREAKERS

As explained on page 34, fuses are manufactured in several types—glass capsule, Autofuse and ceramic. Other circuit-protection devices include fusible links and circuit breakers.

Fuses, regardless of type, are shown as two dots joined by an S-shaped line. This symbol is difficult to draw freehand, so if neatness is important, buy a template of standard electrical symbols at an art- or drafting-supply store. Write the fuse's amperage rating next to the symbol.

Fusible links are drawn as an ordinary wire joined to the protected circuit by a black dot at each end. It is customary to write fusible link, along with gage size, next to the symbol.

Circuit breakers are drawn as a pair of dots joined by a C-shaped line. Usually, there is also a rectangular box around the symbol to portray the circuit-breaker case. Write the amperage rating next to the symbol.

Fuse panels are represented two ways: In a basic-power diagram, the panel is drawn in its entirety. See example, page 90. For an individual circuit, only the fuse(s) applicable to that circuit need to be shown. In this case, indicate the fuse panel as a *phantom component*, using a dotted line.

CONNECTORS & TERMINALS

Connectors and terminals are used to connect wires together or plug them into a component. In mass production, this is preferable to soldering each wire connection separately. It also makes service easier at a later date. Project cars also use terminals because many components to which the wires attach are usually designed for production cars and have built-in terminals.

Because most terminals in use today are of the male/female type, the symbols for terminals are designed to represent

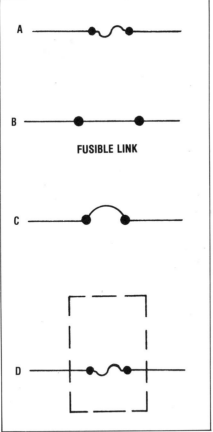

FUSIBLE LINK

A. Fuses are represented by S-shaped line between two dots. Use this to indicate all types of fuses, anywhere in circuit or in fuse panel.
B. Represent fusible link as conventional wire between two dots. Include the label FUSIBLE LINK.
C. Circuit breaker is represented by a semicircle between two dots. In some books, you may find the letter C or CB next to the symbol, but this is unneccessary.
D. Represent fuse panel as a rectangle containing one fuse symbol for each fuse contained in it. Use dotted lines to depict fuse panel in individual circuit drawing. If there are circuit breakers in panel, use appropriate symbol for them.

this distinction. The male-terminal symbol is shaped like an *arrowhead* and represents the *blade* of the connector. The female-terminal symbol is shaped like a *reverse arrowhead,* or *V,* and represents the *socket* the male blade plugs into.

In factory harnesses, there are two types of connectors in common use: the *flat-blade* and *bullet* (pin) type. Some of

the latest Detroit harnesses are also sporting highly weatherproofed connectors for electronic engine controls. These connectors have very delicate pins. General-purpose solderless connectors used for project cars are discussed on pages 63-65. Whether you're using general-purpose terminals or factory ones salvaged from a donor car, the symbols for male and female terminals are the same—arrowhead and reverse-arrowhead.

In the case of wire-to-wire connections, the hot side is female while the side being fed is male. This is because the female terminal is usually built into an insulator block to prevent short circuits if the connection is disconnected, either intentionally or accidentally.

Generally, when several wires attach to the same component, the terminals on factory wiring harnesses have molded insulators containing a separate cavity for each wire. In most cases, there's a female terminal inside each cavity, regardless of whether it's a hot wire or a ground wire.

The mating terminals on components are male blades. Again, the female (hot) terminal within the insulated connector is protected against shorting when the connector is disconnected. The exposed male-blade terminals on the component are in no danger of shorting because there's no voltage in a component when the terminal connector is unplugged.

When designing and drawing your circuit diagrams, pay particular attention to the orientation of male and female terminals used to connect the components you've chosen. Just remember that the hot side of any connection should always have a female terminal.

Polarity-Protected Connectors— These can be plugged together only one way, for the purpose of preventing the wrong wire from being connected to the wrong terminal. Several methods may be employed to provide the one-way feature. One method is to provide a keyway in one connector block with a matching key on the other. When the two connector halves are correctly aligned, the key fits into the keyway. Another method is to group terminals in non-symmetrical arrangements or at angles to each other. Still another method is to mix both male and female terminals arranged in a non-

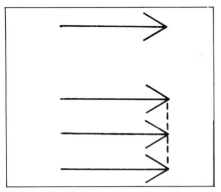

Male terminal symbol is an arrowhead at end of wire. If there are multiple male terminals in a single connector, represent this by adding dotted line to join all terminals.

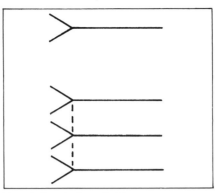

Female terminal symbol is a reverse arrowhead at end of wire. As with male terminals, dotted line is used to represent multiple female terminals in single connector.

JUNCTIONS

Occasionally, a single feed wire is used to supply two separate components. A *junction* is used where the separate circuits branch out from the feed wire. In factory cars, this may be a soldered or crimped joint inside the harness. For project cars, I prefer using a junction block. The electrical path is the same in either case, but a junction block is easier to install. Regardless of the attachment method, use a single black dot to indicate a junction. If you're planning to use a junction block, write the words next to the dot.

SWITCHES

As explained on page 30, various types of switches by the number of *poles* and *throws*. These distinctions should become clear when you draw the switches into your circuit diagrams.

Poles are represented by a straight line appearing to "pivot" from a large black dot. Depending on how many throws the switch has, the pole's free end will be pointing at one dot—or one of several dots—which represent the contacts, or throws. Carefully study the accompanying drawings to see how the pole is drawn. Although it is essentially a straight line, it isn't easily confused with wires, terminals or the like.

Multiple-pole switches are represented by a set of parallel lines, each representing one path for current to flow through the switch. In most cases, the poles are mechanically linked together, so moving

symmetrical pattern in one connector, with mating terminals in the other connector.

Connectors using blade terminals usually confine both male and female terminals within the connector block. If bullet terminals are used, the exposed male terminal should not be hot, for short-circuit protection. If all the wires in a connector must be hot, use one of the other polarity-protection methods.

If you're adapting existing factory-harness connectors to your project car, the previous few paragraphs are relevant to you. But if you're building a harness entirely from scratch, using individual connector terminals, you must use other methods to make the connections foolproof.

When joining several wires to several others in a circuit, you can cut the wire ends in varying lengths, so a long wire on one half of the harness matches with a short wire in the other half, and vice-versa. Or, use color-coded wire connectors of the same size with different insulator colors. Use a male and female of the same color for a matching set of wires. Another way is to use a different wire color for each matching pair in the harness.

As with other terminals, wiring-diagram symbols for polarity-protected terminals must be drawn with the male and female symbols correctly oriented to ensure that you won't accidentally design a circuit with a potential for a short circuit if the wires are disconnected.

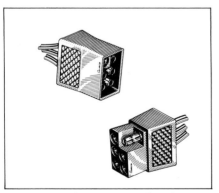

Keyway terminals are designed to go together only one way. The one shown has a raised portion (key) in one connector and a matching depression (keyway) in the other. Designs vary. Drawing courtesy of Au-ve-co Products.

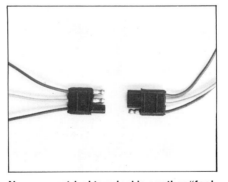

Non-symmetrical terminal is another "foolproof" connection. In this example, connector at left contains one female terminal among two males; mating one has one male among two females. Four-wire version of this is often used for trailer connections. Photo by Jim Barrett.

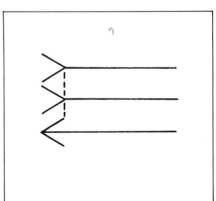

When representing non-symmetrical connection, be sure to draw males and females in correct sequence. Where possible, keep females on feed side of connection to prevent shorts when connection is unplugged.

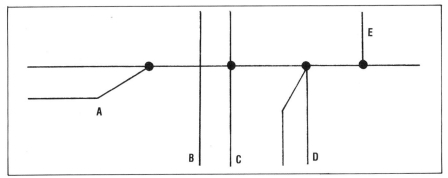

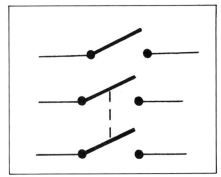

Permanent junctions in wires, either by soldering or with various solderless connectors, are represented by a single dot. (A) shows single wire branching from main line and continuing parallel to harness. (B) shows two wires crossing *without joining,* indicated by absence of dot. (C) shows two wires branching from main line and going in opposite directions. (D) shows two wires branching from main line and going in the same direction. (E) shows a single wire branching from the main line and leaving harness.

Simple switch (top) is represented by break in wire, with two dots. The switch *pole* is the "movable" line between the dots, exaggerated here for emphasis. In multiple-pole switch, poles are connected by dotted line. Double pole switch is shown.

the control handle causes all poles to move simultaneously from one throw to another. The mechanical link is represented by a dotted line joining the related pole lines.

Throws are represented by dots. There's one dot for each position the switch can be moved to, including OFF. *Multiple-throw* switches have more dots than *single-throw* switches, simply because they can be moved to more positions. Regardless of the number of throws, each position provides a current path associated with each dot. The one exception, of course, is OFF, which is a dead-end dot with no current path leading from it.

You may not be able to tell by looking at the outside of a switch how many poles it has, but you can move the control handle to all the possible positions to count

the throws. With a great deal of patience, it's possible to trace the continuity between the various switch terminals in each throw position, and in the process, determine how many poles are inside to make the various continuities possible. Fortunately, this complicated process isn't often necessary because you can usually find a wiring diagram for the vehicle the switch came from, which will show you the pole arrangement. In the case of aftermarket switches, the manufacturer can supply internal-circuit drawings if they aren't provided with the switch when you buy it.

LAMPS

Lamps are symbolically represented by circles. Inside the circle, you'll find a curlicue representing the filament. Some lamps, such as high/low-beam head-

lamps, have two filaments inside one glass envelope. In cases like this, there is one curlicue for each filament. Often, the brighter of the two filaments is represented by a larger curlicue, though this is not always the case. In symbol language, symbol size is not necessarily proportional to actual lamp size.

In a simplified circuit drawing, it's common to show the bulb with the wiring terminals "sprouting" from the circle, although the actual component usually has a separate lamp socket.

RELAYS

The symbol for a relay—or solenoid, depending on function—is based on a rectangle. The magnetic coil is represented by several curlicues or loops. Contact points are represented exactly like those for a switch. The mechanical link between the coil and the points is represented by a dotted line.

There are multiple-pole and multiple-throw relays available, and the symbology for them is exactly like multiple poles and throws in switches. In these cases, the dotted line is drawn to intersect all the poles and the coil.

MOTORS

These may be represented several ways, depending on how much internal circuitry you need to show.

A simple *single-speed motor* is represented by a circle with a capital M inside.

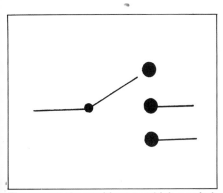

Throws are the positions to which a switch may be moved, including OFF. This is a single-pole, triple-throw switch. Throws are represented by dots, exaggerated here for emphasis.

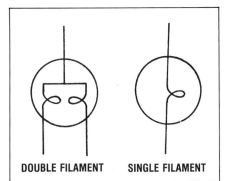

DOUBLE FILAMENT   **SINGLE FILAMENT**

A lamp is represented by a circle, with a separate loop for each filament it contains. Both filaments of double filament lamp share the same ground, which is typical of most lamps of this type.

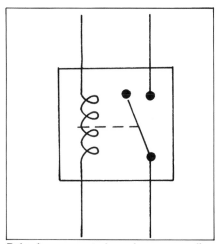

Relay is represented as a box surrounding a coil and switch. Dotted line implies that coil operates switch, which is exactly how a relay works.

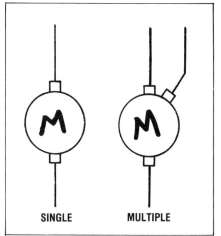

SINGLE   MULTIPLE

Left: Single-speed motor is represented by circle, feed wire and ground. Squares represent brushes. To prevent confusing it with lamp in small diagrams, letter M is written inside circle. Right: Two-speed motor is shown with three brushes, represented by squares.

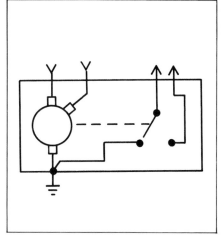

Wiper motor symbol includes park switch. Dotted line implies that switch is operated by motor, which is the case.

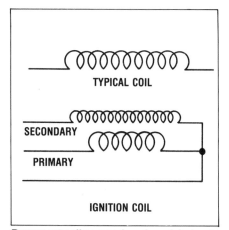

TYPICAL COIL

SECONDARY

PRIMARY

IGNITION COIL

Represent coil as a series of semicircles. In ignition coil symbol, number of turns and relative sizes are indicated.

STARTER MOTOR DIAGRAM

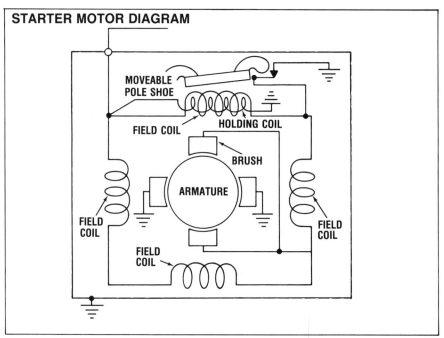

MOVEABLE POLE SHOE
FIELD COIL
HOLDING COIL
BRUSH
ARMATURE
FIELD COIL
FIELD COIL
FIELD COIL

This starter motor drawing is typical of complicated motor. It contains four field coil windings, a pull-in coil, as well as armature. This symbol is actually more of a self-contained schematic. Drawing courtesy of American Motors Corp.

Show the two brushes as small rectangles. A *permanent-magnet, two-speed motor* requires an additional brush, but otherwise is the same.

Windshield-wiper motors usually have a built-in park switch, represented as a SPST switch joined by a dotted line to the motor.

To correctly portray the workings of a complicated motor, such as a starter motor and related hardware, you may want to show internal wiring. The field coils are shown as multiple wire loops, the brushes as small rectangles, and the armature as a circle.

COILS

As mentioned in the discussion on relays and motors, coils are shown as multiple loops. This is true regardless of whether the coil is used as a relay or solenoid, or as a transformer in the ignition coil. In an ignition coil, draw the primary winding with just a few turns and the secondary winding with many more. As with fuses, consider using a template for neatness.

PHANTOM COMPONENTS

I use this arbitrary term to cover a special condition often needed in drawing circuit diagrams. Suppose you're drawing the circuit for your brake lamps. You'll learn later that the brake-lamp circuit is fed by a fuse mounted in a fuse panel. And, you probably already know that a fuse panel contains six or more fuses. In the brake-lamp circuit, only one

fuse is involved, so represent the fuse panel as a *phantom component* to indicate that there's more to the component than is shown in the circuit you're presently looking at. The brake-lamp fuse would, of course, be represented as a solid-line component. The symbol for a phantom component is exactly like the symbol for a non-phantom one, except the *outline* of the phantom is drawn with *dotted lines*.

Components shown in phantom are those that are not actually part of the circuit, but need to be shown for reference. Typical examples are multi-pole switches, the fuse panel and the battery. Also, you occasionally must show a wire that's actually a part of some other circuit in order to indicate its relationship to the circuit you're designing. For example, the wire that brings power from the battery to the fuse block is actually a part of the *basic power circuit*. But, you may want to show it as a *phantom wire* in the brake-lamp circuit to indicate that it is the power source for that circuit. If this point is not completely clear to you now, it should be when you study the specific circuits described later in this chapter.

DRAWING DIAGRAMS

Before you start drawing circuit diagrams, decide which components will be on each circuit. First, make a list of all the *electrical loads* you plan to use in your car—every lamp, gauge, motor and

accessory. Also list the *control device*— switch, rheostat or relay—for each load.

Now, go back and organize the list of components into systems. If your mind works like mine, you'll have groupings like: brake lamps and brake switch; headlamps and headlamp switch; horn, relay and horn switch, and so on.

Next, make a similar list of the components that supply electricity to operate these loads. Here's where you should include the battery, generator, regulator and fuse panel, to name the most prominent ones. This list will eventually become the *basic power circuit*.

The following is a list of the various circuits found in a typical car. Each circuit listed is described in detail in this chapter. Use this list for your car, or adapt it to suit your specifications. Either way is OK, as long as you account for every load, control and supply.

The circuit list looks like this:
1. Basic power
2. Charging
3. Starting
4. Ignition
5. Instruments
6. Windshield wiper and washer
7. Headlamps
8. Parking, side-marker and taillamps
9. Turn, emergency and brake lamps
10. Back-up lamps
12. Horn
13. Miscellaneous circuits: cigarette lighter, courtesy lamps, heater, radio and other accessories.

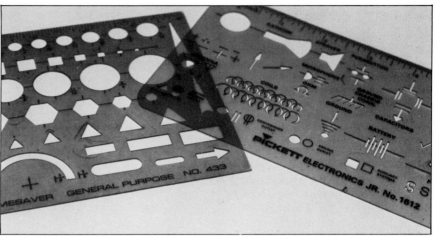

PHANTOM FUSE PANEL

PHANTOM SWITCH

When showing component as reference in circuit, draw it as a *phantom component*. This implies that full component is shown in its completeness somewhere else. Phantom symbol is a dotted line. Shown here are phantom fuse panel and phantom switch.

I use several plastic templates for neatness. You can draw wiring symbols without them, of course. Templates are not expensive and are available at drafting supply stores and some larger office and stationery suppliers.

In the following pages, you'll find two sets of drawings for each circuit listed above. One is for a typical rear-engine car—an air-cooled VW is shown—and one is for a typical front-engine car—a Ford Pinto is shown. These cars were chosen because they are the ones most often used for kit cars. However, the circuit diagrams can be used as general guidelines for designing circuits for any project car. Use them in conjunction with factory wiring diagrams and wiring instructions that come with the components you buy.

The components shown in these diagrams are the ones used in the VW and Pinto factory installations. The diagrams also generally follow the factory wiring scheme, but the circuits are simplified to make them easier to understand for assembling the wiring harnesses for your car.

In simplifying the circuits, the electrical relationship of the various wires and components has not been changed. But, the drawings show routings you'd most likely use in a one-of-a-kind car.

Often, factory diagrams appear incredibly complicated and sometimes seem to defy logic. This is because the factory harness is designed to be easily and unerringly installed on a fast-moving assembly line by workers who may or may not understand what each wire is for. When you wire your project car from scratch, you should assemble your harness with a great deal of understanding and take nothing for granted.

The sample diagrams shown are for cars that were built before the onrush of electronic engine and emission controls. As mentioned earlier, installing such controls in a project car is beyond the scope of this book.

One last reminder before you start drawing diagrams. Each component should be shown as a solid outline *only in the drawing where it's featured*. If the same component or wire appears as a reference in other drawings, it should always be shown as a *phantom component*. This allows you to use your circuit diagrams as a double-check when you assemble the wiring harnesses and connect them to the components. What you'll be doing is marking off each solid-line component or wire on the diagram as

you install it, because it will appear as a solid line on only one drawing. When everything is installed, scan the drawings. Every solid-line component should be marked. And, of course, the phantom components should not be marked. This eliminates the possibility of omitting or duplicating anything.

GUIDELINES FOR DRAWING A DIAGRAM

I followed these guidelines in preparing sample diagrams for this book. If you draw your own diagrams—and you certainly should—also follow these guidelines.

● Before current can flow, each circuit must be complete, from battery positive terminal to the battery negative terminal. Short circuits prevent current from getting to the load, and they blow fuses. Be sure you don't accidentally draw a short circuit.

● The car's engine, frame and metal body act as grounds, or the return path of the circuit. Use a ground symbol to indicate where the circuit grounds to engine, frame or metal body.

● Components mounted to fiberglass need a separate ground wire leading to the frame. Indicate this in the diagrams.

● A switch must be wired in series with the load it controls.

● Loads are usually wired in parallel with each other.

● Be sure each potentially hot wire terminates in a female connector to prevent shorts if accidentally disconnected. A good way to remember this is to make sure the arrow-shaped symbols for both male and female terminals point *toward* the power source.

Electrical & Geographic Considerations—This chapter deals with a concept you *must* understand in order to progress to the next step of your wiring task. It was hinted at in the discussion on the differences between how a factory harness is wired and how you'll wire your project car. It has to do with the fact that the electrical system in your car has two distinctly different sets of characteristics that must be temporarily separated in your thinking. They are *electrical considerations* and *geographic considerations*. Here, the term *geographic* refers to the physical positioning of the electrical components contained in the space within the car's body.

Circuit design deals primarily with the electrical considerations. Geographic considerations are covered in the following chapter. For now, here are some simple examples that will help you differentiate between the two.

The illustration below shows four ex-

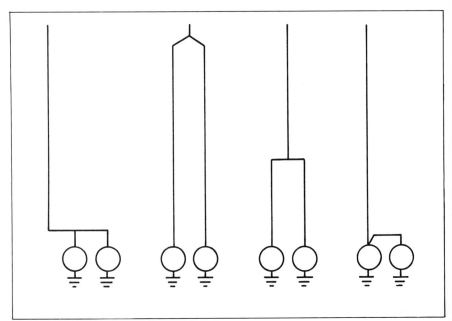

Electrical considerations: These drawings demonstrate four different ways to represent the same thing—two lamps in parallel—as described in accompanying text.

88

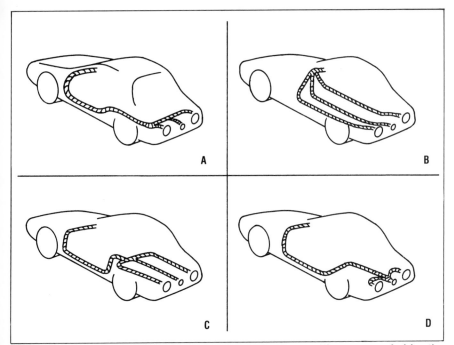

Geographic considerations: These four drawings demonstrate different ways of wiring the tail and license lamps in a car. Although all four methods are electrically sound, method A is the most practical from a geographic standpoint. See text below.

amples of how two identical lamps can be wired. Electrical considerations of these diagrams are identical—that is, the two lamps are wired in parallel. Where the diagrams differ is in their geographic positioning. Routing the wires can take on a wide variety of forms, depending on the geography of the car where they will be installed. Let's look at them.

The illustration above shows the circuit for three lamps wired in parallel. These could be the left and right taillamps, and license-plate lamp. The only electrical consideration in this kind of circuit design is that each lamp be supplied with 12 volts and have a path to ground. Notice that there are several ways to portray this, all of them electrically sound.

But the geographic consideration—how the components relate to each other within a space—for each example is evident when you start to construct the harness. The second illustration shows it is possible to wire a car following each of the circuit drawings exactly. But notice that methods B and C are impractical. They waste a lot of wire. Method A is probably the most common and prefer-

able installation.

So, these exercises demonstrate that it's possible to design the same circuit in several different ways, all *electrically* correct. However, only one may be *geographically practical* for your car. You've probably seen that the way a circuit is drawn on a schematic may or may not resemble the appearance of the harness that's eventually made. Please get this notion firmly planted in your mind. If you don't, I can practically guarantee your wiring project will be confusing and frustrating rather than rewarding and fun.

BASIC POWER CIRCUIT

Battery voltage or generator output is referred to as $B+$. This is the electricity in the wiring system that does the work. You must provide a circuit that can distribute $B+$ to the various loads in the system.

Certain components in a car's electrical system can be regarded as *power centers*. That is, numerous sub-circuits branch out from them. Power centers include the battery, voltage regulator, fuse panel, starter solenoid and ignition switch. These power centers and their

wires form the *basic power circuit*.

There are two kinds of $B+$ circuits—those that are connected directly to the battery (always hot) and those that are controlled by the ignition switch, called *ignition* or *controlled* circuits.

Components that operate without the key turned on are in the always-hot circuits. These include park/taillamps, horn, interior lamps and emergency flashers. In a VW, the radio is usually included because the ignition switch does not have an accessory position. All other components are operated through the ignition switch, which means the key must either be turned to the ON position or the ACC (accessory) position. These shouldn't be left on when the engine is not running.

In the power-circuit drawings, notice that in some cases several fuses in the fuse panel are grouped together. Each group is connected by a *bus bar* at the input. This means a single input wire can supply voltage to more than one fuse, simplifying wiring. For example, in the VW circuit drawing, fuses 1 and 2 are bussed together. In the Pinto circuit, fuses 3 and 4 are similarly bussed.

Notice also that most of the wires carrying current into the fuse panel are solid, while those carrying current out of the fuse panel are shown in phantom. This is because the phantom wires are actually parts of other circuits, and are shown on the power-circuit drawing only as references.

The fuses in the basic power circuit are numbered to correspond to the cavities in the fuse block. Each fuse is identified with the same number in each diagram where it is shown.

The Pinto starter relay and VW voltage regulator are shown in phantom in the power-circuit drawings because, although their terminals act as power-distribution points, the components are actually parts of other circuits.

CHARGING-SYSTEM CIRCUIT

Although the drawings for VW and Pinto charging systems show a number of components, most of them are drawn in phantom. The only ones that are part of the charging system are the generator, voltage regulator, charge-indicator lamp, and optional ammeter and volt-

Basic Power Circuit—VW

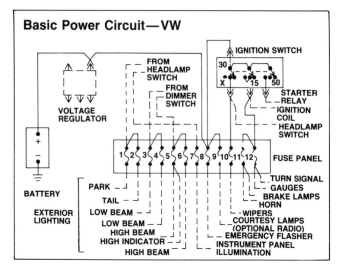

Basic Power Circuit—Pinto

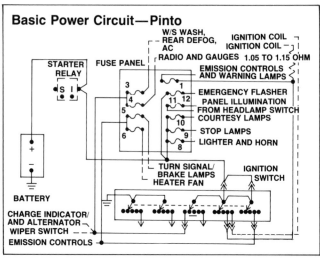

Charging Circuit—VW

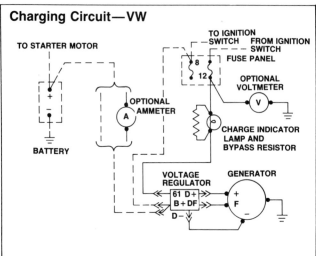

Charging Circuit—Pinto

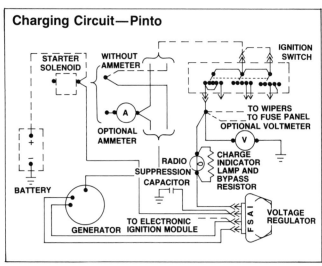

meter. You can add these optional gauges for additional driving information.

In use, the charge-indicator lamp gets B+ voltage from the gauge fuse (VW) or ignition switch (Pinto) and grounds through the voltage regulator. Providing initial voltage to energize the generator like this is referred to as *excitation*. When this happens, the lamp lights, indicating no generator output. When the engine starts to drive the generator and output increases, it supplies B+ to what had been the ground for the indicator lamp. With equal voltage on both sides, the lamp goes out, indicating that the generator is generating voltage. Notice the resistor wired in parallel with the lamp; it provides voltage to excite the generator if the lamp burns out.

When the starter is operated, current stored in the battery is used to run the starter and ignition systems. After the engine starts and the generator is operating correctly, enough current is generated to operate all of the car's electrical components. Unless the load is excessive, the generator produces enough extra current to recharge the battery.

Both the VW and Pinto systems shown use an *external voltage regulator*. Some of the latest generators use a built-in solid-state regulator, called an *integral regulator*, which simplifies the wiring. If your car has an integral regulator, refer to the factory diagram for the necessary wiring. Other than simplified wiring, the generally claimed benefit of an integral regulator is its solid-state electronics, which eliminate mechanical devices that

are used in an external regulator. Although I wouldn't reject a charging system solely on the basis of the regulator type, I prefer the electronic regulator, which has proven extremely reliable.

Optional Ammeter—In both VW and Pinto charging systems, two alternate sets of wiring are provided—one showing the optional ammeter and one showing the same circuit without this gauge. Note the use of the brackets in the drawings above. This is a common way to illustrate optional wiring. When you construct a harness from such a diagram, follow *one* of the alternate paths, not *both!*

The ammeter is connected so the current from the battery—except to the starter motor—flows through it. And, all the current from the generator to the bat-

Starter Circuit—VW

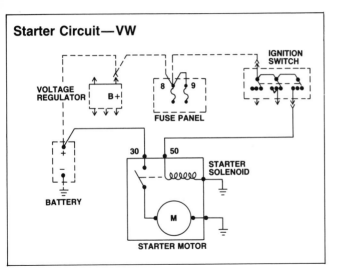

Starter Circuit—Pinto

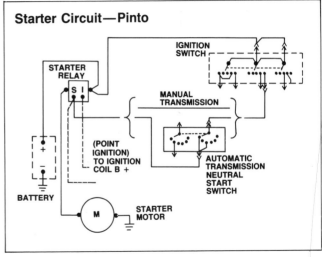

tery also flows through the ammeter. The gauge is calibrated with a + (charge) and a − (discharge) to indicate current flow into or out of the battery. A serious shortcoming of using an ammeter is that you must run two additional heavy wires from the generator to the instrument panel.

Optional Voltmeter—Both diagrams also show an optional voltmeter. The voltmeter has two terminals. One is grounded and the other receives B + through the ignition switch. Essentially, this means the voltmeter reads across the car's entire electrical system. With the key turned on and the engine not running, it reads battery voltage.

STARTER CIRCUIT

The starter circuit provides voltage directly from the battery to the starter motor. Because it's not desirable to run a heavy battery cable to the ignition switch and back, the starter circuit uses a starter relay.

The Pinto system uses a starter *relay* mounted to a convenient under-hood panel. It is a common mistake to refer to this Pinto component as a solenoid. Because it acts only as a relay, it is correctly called a relay. On the VW, the starter *solenoid* is mounted directly on the starter motor. The term solenoid is correctly used here. The device not only acts as an

electrical relay, but also serves to move the starter gear into engagement with the engine ring gear. Regardless of the engine you're using, when the ignition switch is turned to the START position, voltage is sent to the starter relay. The relay coil is energized, pulling the movable element. This completes the circuit between the battery and the starter motor through the short, heavy cables. Cranking takes place as long as the key is held in the START position.

Pinto Starter-Circuit Exceptions—The diagram shows two alternate paths for the current from the ignition switch to the starter circuit. On cars with a *manual transmission,* the path is unbroken to the starter relay. On cars with an *automatic transmission,* a safety device called a *neutral-start switch* prevents starter operation except when the transmission selector is in PARK or N (neutral). Many new cars equipped with manual transmissions also use a similar switch on the clutch pedal.

On Pintos with conventional point ignition, the starter relay has a special terminal marked I. The starter relay on Pintos with electronic ignition has two wires attached to the S terminal of the starter relay. These are explained in the following discussion on ignition circuits.

On the Pinto, the battery terminal on the starter relay is also used as the B +

distribution point for all electrical systems in the car. You could also use this wiring setup on the VW, although the factory method usually provides two cables at the battery + post—one for the starter and one for the rest of the car. Refer to the power-circuit drawings for details.

IGNITION CIRCUIT

The VW and Pinto circuits are significantly different, so they are covered separately. Both diagrams show the connection point for an optional tachometer at the ignition coil's negative terminal. Refer to the instruments diagram on page 93 and the tachometer manufacturer's instruction sheet for further information.

VW Ignition System—At the right of the diagram, I show two possible ways to wire the ignition feed to the coil. In the top drawing, the ignition switch has two wires attached, one to the fuse panel and one directly to the ignition coil. In the bottom drawing, the ignition switch feeds the fuse panel only. Ignition feed is provided by the bus bar in the fuse panel. Both methods are electrically identical, and there's nothing to recommend one over the other. Depending on where you install various components, one or the other may make better sense in your specific car.

The VW system uses a set of points

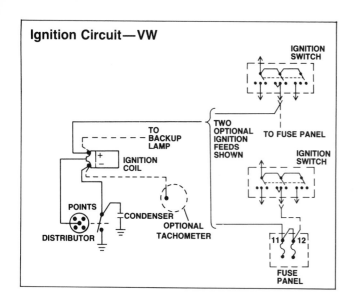

Ignition Circuit—VW

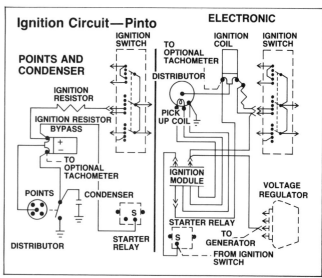

Ignition Circuit—Pinto

ELECTRONIC

POINTS AND CONDENSER

and a condenser. When the ignition switch is in the START or RUN position, it provides voltage to the + terminal of the *ignition coil*. Ground for the coil is provided through cam-operated *ignition points* in the distributor. As described earlier, with the points closed, a magnetic field builds up in the coil primary winding. As the points open, the field collapses and high voltage is induced into the secondary winding. It flows to the distributor and on to the sparkplugs.

Note that the movable point is insulated and is connected to a *condenser,* which acts as an electrical cushion when the points open. This reduces point arcing.

As far as the wiring harness is concerned, the ignition circuit consists of only one wire between the ignition switch and the coil + terminal. For our purposes, the high-voltage wires are regarded as part of the engine. On VWs, the coil + terminal acts as a distribution point for B+ current to the back-up lamps and some other engine components. The installation of these wires is covered in their individual circuits.

Pinto Point-Ignition System—This system was used only for the first few years of manufacture. It is essentially like the VW system already explained. The only significant difference is that a resistor is wired between the ignition

switch and the coil + terminal. This reduces voltage to preserve the ignition points. During engine start-up, the starter relay provides full voltage to the coil through the I terminal on the relay body. This provides maximum spark for good starting characteristics. If you forget to include the resistor in your harness, point wear will be excessive. If you forget the wire from the I terminal, ignition voltage during engine start-up will be reduced.

Pinto Electronic-Ignition System—Although this book does not cover electronic components in general, electronic ignition is so common today that it is included here. Although this is a specific discussion of the Pinto system, various other electronic-ignition systems operate, and are wired, in essentially the same way. Refer to the appropriate shop manual for wiring details.

Current for the system starts at the ignition switch, which powers up the electronic-ignition module and voltage regulator any time the ignition switch is on. Another branch of this same circuit carries current through a resistor to the coil + terminal when the ignition switch is on. Ground for the coil is provided through the module. The module opens and closes the ignition-coil ground, obeying instructions from a pickup coil in the distributor. The pickup coil and module essentially replace the con-

ventional points and condenser.

When the engine is running, the coil receives reduced voltage through the resistor. During engine start-up, the coil receives full voltage from the S terminal of the ignition switch. Simultaneously, a start signal is sent from the starter-relay S terminal to the ignition module.

When drawing and assembling an electronic-ignition circuit, be sure every wire is connected to each component *exactly* as it was in the factory installation. Electronic components are not forgiving of mistakes. If you're using some other brand of electronic ignition, obtain a shop manual for the car to see how the system is wired, then copy it exactly.

INSTRUMENTS

Both the VW and Pinto—in fact, just about any car you may use as the donor for your project car—rely on minimum instrumentation from the factory—usually a fuel gauge and speedometer. Some, like the Pinto, offer factory add-on instruments—often called a *rally package*. These instruments are usually not esthetically adaptable to most project cars. In fact, the Pinto uses a *constant voltage regulator (CVR)* to limit the amount of voltage sent to the gauges. This, in turn, requires a radio-noise suppressor. Take my word for it—you probably won't want to bother with this type

Instrument Circuit—VW and Pinto

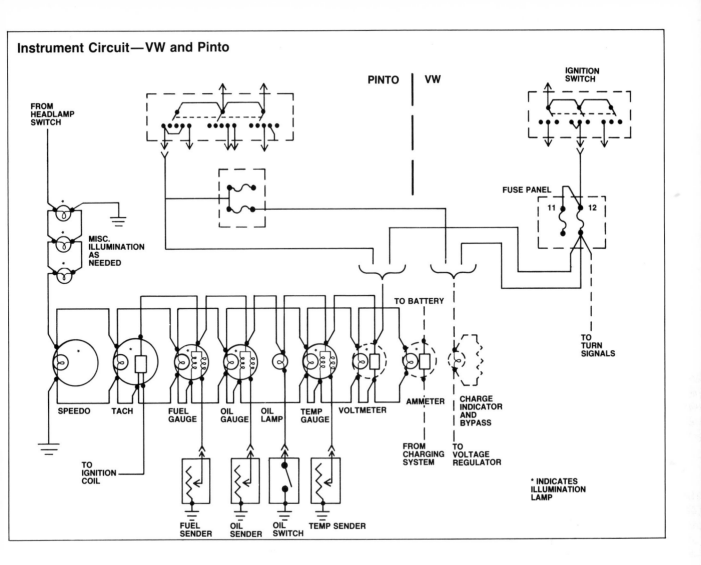

of setup. To minimize your temptation, the Pinto rally package is not even shown. I recommend you use high-quality aftermarket magnetic gauges instead. Manufacturers of high-quality gauges include Stewart Warner, VDO, Sun and Auto Meter. These firms are listed in the supplier's list.

Wiring diagrams here show the use of magnetic gauges, both for the VW and Pinto. They do, however, use the factory ignition switch and fuse-panel hookups, as far as possible. Note the differences between the Pinto and VW ignition switches and fuse panels, shown at the top of each diagram. Also included are references to the charge-indicator lamp and bypass resistor, and an optional ammeter and voltmeter, although these

components are actually a part of the charging system, so are covered elsewhere. To simplify the diagrams, the turn-signal indicator lamps and the high-beam indicator lamp are not included. These are covered in their respective systems, pages 94 and 96-97.

Because the instrumentation on your car will probably be unique, the diagrams do not show the gauges and lamps arranged in any specific order. Instead, they show a typical hookup that is electrically correct. You must adapt the wiring to fit the component arrangement in your car. In many cases, it is possible to use a *continuous loop* of wire to carry B+ to all the gauges, another loop for instrument-panel illumination and a third for ground. However, an individual cir-

cuit must be provided for each *sending unit*.

Also note that the diagram above includes both a gauge and a warning lamp for oil pressure. If you plan to include both displays in your car, be sure to use the correct sending unit for the gauge and the correct switch for the warning lamp. They are not interchangeable.

The diagram also includes both an optional ammeter and voltmeter in addition to the standard charge-indicator lamp for the charging system. You may elect to use both, or either one separately. Refer to the charging-system diagram on page 90 for the complete ammeter, voltmeter and indicator-lamp hookups.

The factory instrument cluster for the Pinto includes warning lamps for oil

Wiper Circuit—VW

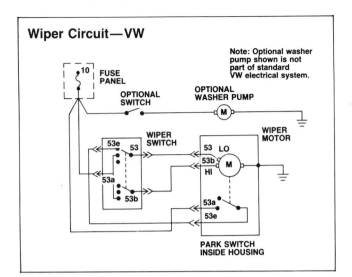

Note: Optional washer pump shown is not part of standard VW electrical system.

Wiper Circuit—Pinto

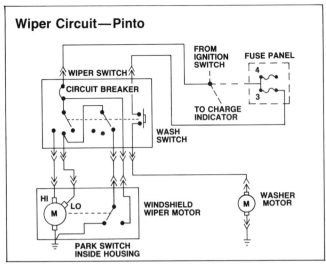

Headlamp Circuit—VW

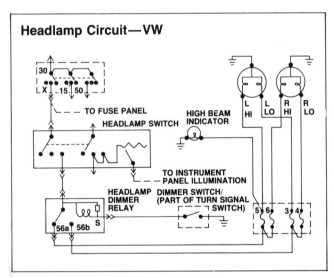

Headlamp Circuit—Pinto

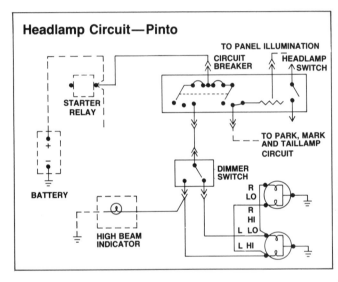

pressure and coolant temperature. These lamps are grounded by senders when the appropriate system is in the danger range. Factory installations are wired so you can tell whether or not the lamp bulbs are burned out. Each time the ignition switch is turned to the START position, a temporary ground is provided for the warning lamps, lighting them.

WINDSHIELD WIPER & WASHER

Both of the systems shown use a two-speed, permanent-magnet motor to operate the wipers. The Pinto uses a one-speed motor to pressurize the washer water, while the VW relies on a non-electrical air-pressure system operated by the spare tire. For convenience and reliability, use a Pinto-type pump system in VW project cars.

Another difference is that the VW system uses a fuse located in the fuse panel to protect the wiper-motor circuit, while the Pinto wiper switch contains a built-in circuit breaker. Specific circuits follow.

The VW system provides B+ to the park switch and to a busbar in the wiper switch. When the switch is turned off and the motor is still running, the park switch provides voltage through terminal 53e of the wiper switch. Current flows out of the switch at 53 and to the low-speed brush of the motor. The motor runs until the park switch opens. In low and high positions, current is applied to terminal 53a of the wiper switch. It flows through terminal 53 for low speed and 53b for high speed.

Taillamp Circuit—VW

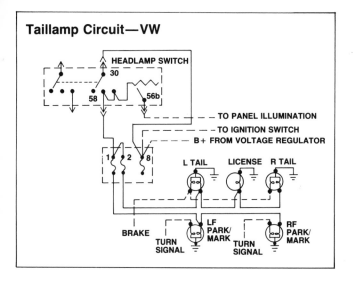

HEADLAMP SWITCH
30
58   56b
— TO PANEL ILLUMINATION
— TO IGNITION SWITCH
— B+ FROM VOLTAGE REGULATOR
1 2 8
L TAIL   LICENSE   R TAIL
BRAKE   TURN SIGNAL   LF PARK/MARK   TURN SIGNAL   RF PARK/MARK

Taillamp Circuit—Pinto

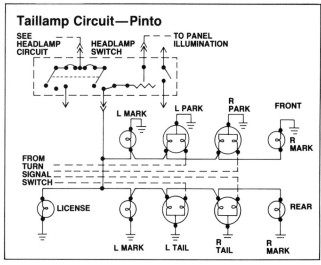

SEE HEADLAMP CIRCUIT   HEADLAMP SWITCH   TO PANEL ILLUMINATION
L MARK   L PARK   R PARK   FRONT   R MARK
FROM TURN SIGNAL SWITCH
LICENSE   REAR
L MARK   L TAIL   R TAIL   R MARK

The wiring diagram for the VW shows a non-standard electric pump for the windshield washer. Because the standard system does not use an electric switch for the washer, you'll need a momentary-on push button for this installation.

The Pinto system provides voltage from the circuit breaker to the park switch and the wiper-control switch. The washer switch gets its power from a fuse in the fuse panel.

When the switch is turned off and the motor is still running, the park switch provides voltage to the wiper switch. From there, current flows through the low-speed brush. When the park switch opens, the motor stops.

In low and high speeds, the circuit breaker provides voltage directly to the switch, which directs it to either the low- or high-speed brush.

The wiper motor is operated by a push button. In the OEM switch, the push button is mechanically linked to the low-speed switch, which causes the wipers to run. The washer pump stops when the button is released, but the wiper motor continues to run until the switch is turned off manually.

LAMPS

This section covers all lamps generally used on a car's exterior—headlamps, taillamps, brake lamps, parking lamps, side-marker lamps, turn, hazard and brake lamps, and back-up lamps.

Headlamps—The VW and Pinto systems operate entirely differently. The VW relies on several fuses to protect the headlamp circuits, while the Pinto uses a circuit breaker in the headlamp switch. The VW uses a control and relay mounted on the steering column to operate high and low beams, while the Pinto uses a foot-operated dimmer switch. Both cars use a single indicator lamp, wired parallel to the high-beam circuit, to indicate that the high beams are on.

Taillamp and marker-lamp circuits are entirely independent of the headlamp circuit on both cars, even though the switches for both are contained in a single switch body and are operated by the same control. Taillamp circuits are described on page 96. Also see diagrams above.

In the VW system, voltage is supplied through the ignition switch to the headlamp switch. This circuit is live only when the ignition switch is in the RUN

position, so the headlamps cannot be left on accidentally. From the headlamp switch, current flows through the relay, which routes current either to high or low beams.

A grounding switch in the turn-signal lever operates the relay. The relay is normally in the high-beam position and moves to the low-beam position when the dimmer switch is closed. The high-beam indicator lamp is wired in parallel with the left high beam. Notice that each headlamp has two fuses, one for each beam.

In the Pinto system, voltage from the battery is always available at the headlamp switch. When the switch is turned on, current flows to the dimmer switch, which routes it to either the low or high beams. The high-beam indicator is wired in parallel with the high beams. A circuit breaker in the headlamp switch protects headlamp wiring.

Incidentally, the Pinto system is typical of most American cars, in that the headlamp circuit uses a circuit breaker instead of fuses. And, it uses a foot-operated mechanical dimmer switch rather than a relay.

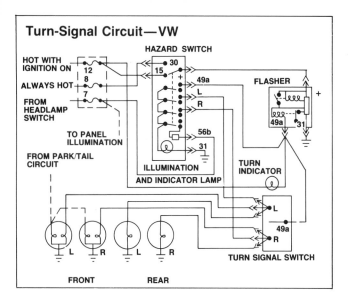

Turn-Signal Circuit—VW

HOT WITH IGNITION ON — 12
ALWAYS HOT — 8
FROM HEADLAMP SWITCH — 7
TO PANEL ILLUMINATION
FROM PARK/TAIL CIRCUIT

HAZARD SWITCH
15 30
49a
L
R
56b
31

ILLUMINATION AND INDICATOR LAMP

FLASHER
49a 31
+

TURN INDICATOR

L
49a
R

TURN SIGNAL SWITCH

L R L R
FRONT   REAR

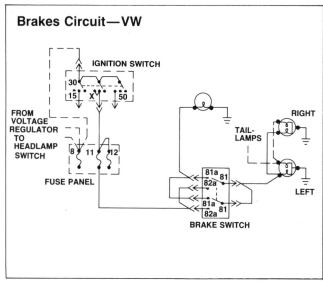

Brakes Circuit—VW

IGNITION SWITCH
30
15 X 50

FROM VOLTAGE REGULATOR TO HEADLAMP SWITCH

8 11 12
FUSE PANEL

81a 81
82a
81a 81
82a
BRAKE SWITCH

RIGHT
TAIL-LAMPS
LEFT

Parking Lamps, Side-Markers & Taillamps—As pointed out in the discussion on headlamps, all of these lamps are controlled by the headlamp switch, even though they are on a separate circuit from the headlamps.

In the VW system, these lamps are operated from fuse-protected circuits. A fuse routes voltage to the headlamp switch, where it is distributed to the instrument-panel illumination and to the parking, side-marker and taillamp circuits. Voltage for the instrument-panel illumination, actually a part of the gauge system, is provided directly to the instrument cluster. Voltage for the taillamps, rear markers and license-plate lamp is routed back to the fuse panel and through a fuse before going on to the lamps. Voltage for the front parking and side-marker lamps is also routed to the fuse panel, through still another fuse.

In the Pinto system, voltage for the parking lamps, side-marker lamps and taillamps is provided by a circuit breaker inside the headlamp switch. This circuit breaker is shared with the headlamps. Two branches—one going to the front of the car and the other going to the rear—carry current to the lamps in this system.

Turn, Hazard & Brake Lamps—The entire system is different on the VW and Pinto. In fact, the VW turn-signal/hazard system is entirely independent of the brake-lamp circuit, while the Pinto brake-lamp circuit is an integral part of the turn-signal circuit. This is because on the VW, the front and rear turn signals are amber, while on the Pinto the front ones are amber, but the rear are red and are shared with the brakes.

In the VW system, three sources of voltage are provided to the hazard switch. They are: *hot with ignition on; always hot,* and *hot with parking lamps on.* Each of these sources includes a separate fuse.

It's easiest to explain the *turn-signal circuit* if we begin with the hazard switch turned off—when it's on, it prevents normal turn-signal operation. When the turn-signal lever is turned to the right or left, current flows through the hazard switch, flasher, turn-signal-lever switch and finally the turn-signal lamps at the side selected. The indicator lamp is not

illuminated. The flasher circuitry times out one flash cycle, then trips the two relays inside the flasher body. This temporarily cuts off current flow to the turn-signal lamps, and turns on the indicator lamp. Then the timer trips the relays again, illuminating the turn lamps and shutting off the indicator lamp. This cycle repeats as long as the turn-signal lever is turned left or right.

The operation of the *hazard circuit* is not affected by the position of the turn-signal lever, because the topmost pole of the hazard switch in the illustration cuts the turn-signal-lever circuit whenever the hazard switch is on. In the ON position, the hazard switch provides a path for current to flow to all four turn-signal lamps simultaneously. It uses the same flasher unit as the turn signals. Each time the timer cycles, the turn-signal lamps go off and the turn-signal-indicator lamp lights.

Also, note the illumination lamp in the hazard switch. If the headlamps or parking lamps are on, the illumination lamp is provided reduced voltage through an electronic module in the hazard switch,

Brake-Lamp Circuit—Pinto

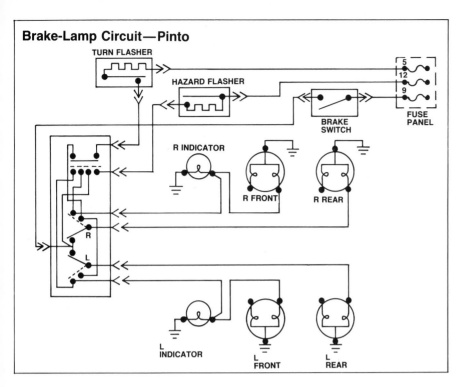

drawn as a square in the circuit diagram. When the hazard switch is turned on, full voltage is provided to the illumination lamp and it glows with full brilliance to indicate hazard-lamp operation.

The VW *brake-lamp system,* shown in the diagram on the facing page, is simple in operation. It gets its voltage from a fuse. When the brake pedal is depressed, the brake switch closes, providing voltage to the brake lamps. If the brake pedal is depressed and one or the other brake circuits is low on fluid, the brake-warning lamp is provided with voltage and it illuminates.

In the Pinto system, three sources of voltage are provided to the turn-signal and hazard circuits. Two of these are always hot, and one is hot with ignition on. Each of these sources includes a separate fuse.

Let's look at *brake operation* first, assuming that the turn-signal lever is straight ahead (off). Voltage is provided by stepping on the brake pedal, closing the brake switch. From the brake switch, it flows through the turn-signal lever without stopping. It is routed from the turn-signal lever through two wires, one going to each of the rear turn/brake lamps.

If the turn-signal lever is turned to the right, the brake-lamp signal is routed only to the left rear turn/brake lamp, because the right rear turn/brake lamp is flashing on and off. The opposite condition occurs if the turn-signal lever is turned to the left.

Now let's see what happens in the *turn signal* part of the circuit during this operation. As mentioned, when the turn-signal lever is in the straight-ahead position, the turn-signal lamps are off, permitting both brake lamps to operate normally.

When the turn-signal lever is turned to the right, current flows first through the turn flasher, then through the right-turn circuit of the turn-signal lever, and finally from the turn-signal lever through two wires. One wire goes to the right-front turn-signal lamp and the right-turn indicator on the instrument panel, while the other one goes to the right-rear turn/brake lamp.

Current flowing through the flasher

quickly heats the thermal element. The movable contact then breaks the current flow to the lamps, which go out. As soon as the element cools enough, current flow is restored to the lamps and the cycle repeats. Operation of the left-turn circuit is identical.

Flashing side-marker lamps are a special case, not included in either the VW or Pinto wiring diagrams shown. Have you ever noticed that on some cars the front side-marker lamp flashes on and off simultaneously with the turn signal? And, when the parking lamps are on, the side marker flashes alternately with the turn signals? Did you ever wonder why? Here's the answer:

In the diagram on page 98, note how the front side-marker lamps are wired. One side of the filament is attached to the hot side of the parking lamp and the other is attached to the hot side of the turn-signal lamp. Guaranteed not to work, right? Wrong! It works just fine.

Let's use the left marker as an example. In the parking-lamp mode, the marker gets voltage from the parking-lamp circuit and it grounds *through the*

Backup-Lamps Circuit—VW

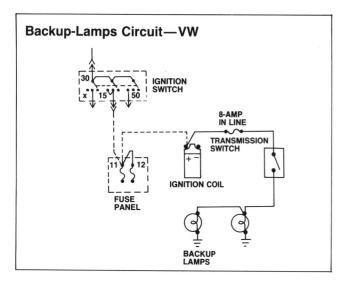

Backup-Lamps Circuit—Pinto

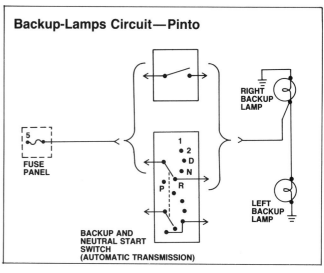

Typical Flashing Side-Marker Circuit

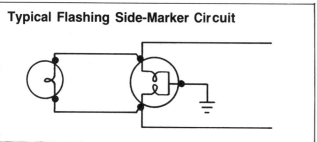

lamp filament of the left-turn signal! You may recall from the earlier discussions of current flow, voltage drops and amperage, that when two resistances are wired in series, the current flow in the circuit is determined by the total resistance in the circuit. If the resulting current flow is too low, it's possible for the feeble current to flow through a load without doing any work. This is exactly what happens with a front turn signal. The side-marker lamp has considerably more resistance than the turn signal, so the side-marker lamp lights, and uses most of the available voltage. The little voltage that's left simply flows through the turn-signal lamp, without lighting it, on its way to ground.

The situation is entirely different, though, when the turn signal is turned on. Each time the turn-signal lamp flashes, the side marker temporarily gets B + current on both sides of the filament, so it goes out. It then continues to flash alternately with the turn signal. That is, the

marker is off when the turn signal is on, and vice-versa.

If the turn signal is operated with the parking lamps off, the side marker gets voltage from the turn-signal circuit and grounds through the parking-lamp filament. In this case, the side marker and turn signal flash simultaneously.

Incidentally, in all cars, the rear side markers are not factory wired to blink with the turn signals like the front ones. If you want them to blink on your project car, wire the rear markers the same as the front ones, attaching one lead to the tail-lamp circuit and one to the turn-signal circuit. But be forewarned: the rear side markers will light up when the brakes are applied in the daytime and will go out when the brakes are applied at night!

Back-up Lamps—In both the VW and Pinto systems, the lamps will work only if the ignition is turned on and the transmission shift lever is in reverse.

In the VW system, the voltage to

operate the back-up lamps is provided by the same wire that feeds the ignition coil. Other than sharing the same feed wire, the ignition and back-up lamps are entirely independent of each other. Note the use of the in-line fuse. This is unusual: Fuses are almost always in the fuse panel. The back-up-lamp system shown here is designed to work only on VWs with manual transmissions—the air-cooled Beetle used for this example is not available with an automatic transmission.

In the Pinto system, there are two transmission options, manual and automatic. Other than the configuration of the switches, the back-up lamps for both transmissions operate exactly the same. Voltage is supplied through the fuse. Current flow is controlled by the position of the back-up-lamp switch. After flowing through the lamps, current is grounded. Note that the back-up-lamp switch used on automatic transmissions is contained in the same switch body as the neutral-start switch, which is covered in the section on the Pinto starter circuit, page 91.

HORN

The horn system, although simple, is important from a safety standpoint. But the horn circuits are shown separately for another reason—they don't seem to fit into any of the other circuits covered so far!

In the VW system, current flows from the fuse block, through the horn, through

Horn Circuit—VW

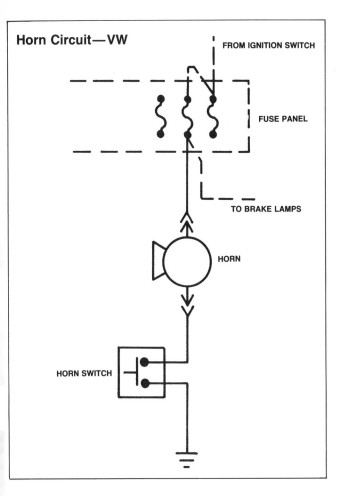

Horn Circuit—Pinto

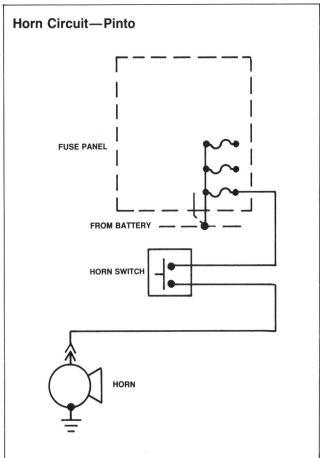

the horn switch, to ground.

In the Pinto system, current flows from the fuse block, through the horn switch, through the horn, to ground.

MISCELLANEOUS CIRCUITS

There are numerous other simple circuits used in factory cars that you may or may not want to incorporate into your project car. These include interior courtesy lamps, cigarette lighter, and in the case of the Pinto, the heater. Most of these circuits are so simple that you should be able to design them without accompanying diagrams, so these circuits are not discussed here.

However, it will be helpful if you understand how the Pinto's heater switch works. As shown in the diagram at right, the switch has four positions—OFF, LOW, MEDIUM, and HIGH. In the OFF position, current cannot flow, of course. But in the low- and medium-speed settings, there are some resistors to pay attention to.

In the low position, current flows through both resistors. This limits amperage in the circuit to a low level and the motor turns slowly. In the medium position, current bypasses one of the resistors, flowing through only one of them. Amperage is somewhat greater than in the low position, so the motor operates at a medium speed. And, in the high position, current flows directly to the motor, so the motor operates in high current flow at maximum speed.

Heater switches in many cars work this way. If the heater motor you're using in your car has the resistor assembly mounted in the heater housing, it's there for a good reason. It is designed to use some of the air flowing through the heater to help cool the resistors. They can become quite hot, especially in the low-speed setting.

Heater Circuit—Pinto

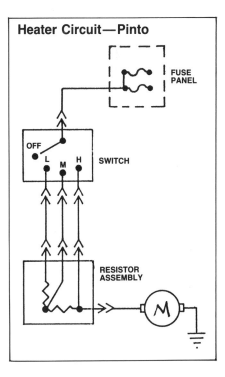

Making a Harness

This well-done wiring job didn't just happen. It was planned. And it followed the procedure I recommend here: On-car planning and on-paper planning, followed by installation according to plan.

If you've completed your circuit designs, you're probably eager to start cutting wires and wrapping them into bundles. But first, devise a plan for cutting and routing all those wires. Otherwise, you may end up with a vinyl-covered copper spiderweb that won't make any sense to you. Doing the job right means correctly translating your circuit designs into *wiring-harness designs*. So, this procedure will be explained first, followed by two different harness-making procedures.

HARNESS DESIGN

Here's where everything you've learned so far comes together—electrical theory, selecting and installing components, and circuit design. When designing circuits, your primary job was to satisfy the *electrical requirements* of the components. That is, you had to make sure each load was provided with a

protection device, control, feed and ground, to name the most common requirements.

In harness design, you must decide how to bundle and route wires to provide the necessary circuits, while at the same time, satisfying certain *geographical and physical requirements*. These include:
- Specific harness routing
- Minimum wire length
- Minimum number of through-the-body grommets
- Minimum number of connectors and terminals
- Wire color and gage
- Neatness

The actual process of harness design involves seven steps. The first three are done on the car; the remaining four on a piece of paper. They are:
- Plan where to run the main harness bundles.
- Plan where branches will leave the

main harness to run to the components.
- Decide where to break the wiring system into smaller, manageable harnesses.
- Sketch a top view of the car, showing where each component is located relative to the others.
- Sketch the routing of the harnesses relative to the components they serve.
- Show where each wire leaves the harness wrap to connect to a terminal.
- Systematically number each wire.

ON-CAR STEPS

Examine several factory-built cars that have the same general wiring configuration as the car you're going to wire—front-engine, rear-engine, mid-engine or whatever. Note how the main harnesses are routed, as well as the smaller ones branching from them. Some of the smaller harnesses may branch again before finally connecting to a component.

There's at least one harness, maybe several, servicing the instrument panel

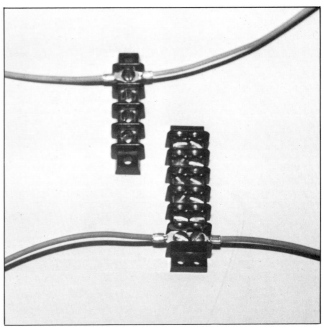

Factory cars use bulkhead connectors as a point for joining harnesses outside body with those inside body. You probaby won't use this exact procedure for your hobby car.

Instead of a bulkhead connector, you'll probably use terminal blocks to join harnesses. Shown here are single-row type (top) and double-row type. Of the two, I prefer the double row. Regardless of type you choose, function is same: to join two or more harnesses together into a system.

and under-dash area, probably terminating at a huge *bulkhead connector*. Outside the passenger compartment, there's a main harness leading from the bulkhead connector to the engine. Note here that if you're wiring your car from scratch, you'll probably be using a terminal block in place of the factory-type bulkhead connector.

Another harness goes to the front lighting, and still another to the rear lighting. Naturally, there will be some differences between cars you look at, but generally the harnesses will fit this description pretty well.

To review, most production-line cars include a large number of options, so harnesses must be easy to install and connect together under assembly-line pressures. As a result, harnesses on a factory car sometimes don't seem to make a lot of sense. Several small harnesses may be plugged together to make one large one, when it would seem to make better sense simply to start with a single harness. That's one advantage you'll have when designing your own harness—simplicity of design.

It helps to have a good imagination because I'm going to ask you to visualize

the harness routes in your car. If you've followed the procedures in the preceding chapters, your car's electrical components are either already installed, or at least were test fitted so you know where they go. Now, picture a harness running through all of the four main areas—instrument panel, front lighting, rear

lighting and engine—with branches going out to the various components.

I have found it helpful to use a visual aid for this step. Use a piece of vinyl-covered clothesline to mock-up each harness. Construction details start on page 105, but here's the basic procedure:

First, cut a piece of clothesline, posi-

My car was this far along when I was ready to begin planning the electrical system. I decided to install rear wiring harness to frame before dropping body into place, because I wanted to attach harness to frame rail for safety.

ROUTING GUIDELINES

While you're deciding where to route the main harness, keep these guidelines in mind:

● Each component must be serviced by the main harness, either directly or by a branch.

● The instrument-panel harness must connect to the front- and rear-lighting harnesses and the engine harness. This junction is the *key connection point* in any wiring job, with the four major harnesses radiating from it.

● The harness is not a circuit. Rather, it is a bundle of wires that form paths for several circuits. Individual circuits branch into and out of the harnesses as necessary.

● Route harnesses around open areas rather than across the middle. Edges usually provide additional protection from damage and also offer more places to tie or clip the harness for best support. Also, the installation usually *looks* better.

● Route harnesses along nonmoving structural members. A good example is the brace that runs from the instrument panel to the bulkhead. It's a natural place to route a harness. So are frame rails, radiator supports, door sills and the driveshaft tunnel.

● Route harnesses away from potential hazards: sharp edges; exhaust pipes; moving parts such as throttle linkage, engine fan, drive shaft and wiper-motor linkage; pinching from hood, trunk and door hinges.

● Generally, wires aren't subjected to much motion or flexing if they're clipped in place. But some flexing is inevitable in certain places. For instance, be sure to allow some slack in the harness between the engine and the chassis. Also, if you're installing door-mounted radio speakers or underhood and trunk lamps, allow for flexing at hinge locations.

● Plan for harness supports. Use a support about every 12 inches unless the harness runs under a floor carpet, which will hold it down.

● At terminal ends, don't depend on terminals to support the harness: Harness weight could cause terminals to disconnect or fail. Support the harness with ties or clips at no more than 6 inches from terminals.

● Keep the wires neatly bundled into harnesses. Use nylon ties, loom tape, large shrink wrap or flexible conduit.

I used clothesline to mock up harnesses inside body. Note branch points for individual gauges sticking out of gauge holes. Each single clothesline branch represents all wires that will go to each gauge: B+, ground, illumination, and in most cases, a sender wire.

tion it where you think the final harness should go, and decide how it looks. No good? Throw it away and try again. It's much cheaper to discard 10 feet of clothesline than 10 feet of finished harness. So take time to plan.

At points where you think smaller branches should run from the main harness, cut lengths of clothesline and tape a piece at each branch point. When you're done, the car will be "wired" with clothesline! Don't let your buddies see this, or they'll think you've cracked! The clothesline "harness" will be used for making the actual harness, so keep it.

After the mock-up harness is in the car, notice how unwieldly it would be if taken out of the car and laid on the floor. So, decide now where to divide the final harness into several smaller, more manageable harnesses. It's up to you, but here's how it is typically done: engine harness, front-lighting harness, rear-lighting harness and instrument-panel harness. Generally, a good place to break a harness is where it must pass through the car body, or at a bulkhead. Later, when you assemble the individual harnesses and install them, they'll be joined together with terminal blocks. So, when you decide where to divide your harnesses, also decide where the terminal blocks will go.

ON-PAPER STEPS

You don't have to be an artist or draftsman to do this next step. But it does help if you get your patience in good working order.

Start with a piece of clean paper at least as large as an open newspaper, preferably larger. Use butcher paper, leftover holiday wrapping, or drawing paper purchased at an art or stationary store. I use chart pads like those sometimes used by speakers to illustrate a speech. They're about 27 in. X 34 in. These are available at larger office-supply stores. One brand to ask for is Ampad Easel Pad No. 150. There's enough paper on one pad to do harness designs for a number of cars.

Fasten the paper to a garage wall close to your project car. Label one short side of the paper FRONT and the other REAR. The long sides, of course, are labeled LEFT (driver side) and RIGHT (passenger side). Try this trick: Orient the paper in the same direction that the car is facing, such as on the garage wall in front of the car, with the FRONT on the drawing up. You'll find this makes it easier to transfer information from the car to the drawing.

Use a black fine-line marker to draw the outline of your car as though you were looking down on it from the top. This is called a *plan view*. It's to show approximately where each electrical

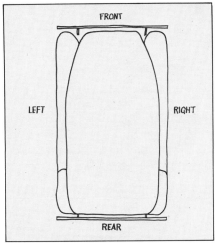

This is the plan view of my car. Notice how I've labeled front, rear, left and right. You don't have to be a Rembrandt to make a simple drawing like this—just make it neat.

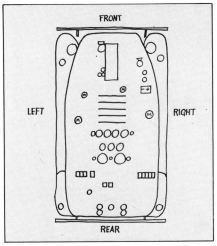

At this point, I've blocked in the major components for reference. Engine is simply a rectangle, lamps and gauges are merely circles. Instrument panel area is drawn disproportionately large because most of car's wires end up here.

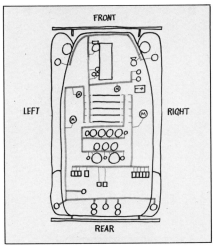

I've indicated harnesses on my plan view. It's a good idea to use a different color for this, such as red. This doesn't show in the black and white photo, however.

component is located in relation to the general shape of the car and to each other. Be sure to include terminal-block locations. It will also show locations of the harnesses and individual wires that connect the components. Remember, be neat and accurate. This drawing will serve as the final guide for wiring your car.

Use simple symbols to depict components—a circle for a light bulb, a rectangle for the engine, and so forth. *Label everything.* When you get done, your drawing will probably resemble the center drawing above. Note that the instrument panel is the largest area because it contains so many electrical components. The drawing doesn't have to be to scale. Just be sure everything is in its correct position and is large enough so you can read it.

Now, referring to your car again, note where you positioned the pieces of clothesline that represent the harnesses. Indicate these on your drawing in a different color, such as red. Draw these main harnesses extra bold so you can see them clearly. Also, be sure the harnesses pass through the terminal blocks. Refer to the drawing above right.

My car is a front-engine, rear-drive, and it has the four main harnesses previously mentioned—instrument panel,

front lighting, rear lighting and engine. Because of the way things worked best, I further separated the instrument harness into a gauge harness and a switch harness. You may or may not need to do so. Just be flexible and do what works best for your car. Also notice in the drawing that all terminal blocks are in one location under the instrument panel.

From your examination of several factory-made cars, you know that each main harness has small branches leading to separate components. Individual wires usually remain inside the branch harness until they get close to the component, where you can then see the individual wire colors and the terminals. If a component has three terminals, you'll find three separate wires coming from the harness at that point. It's the same at each component.

Indicate these individual wires on your drawing, using a third color, such as blue. Draw in each wire separately, showing where each leaves the red harness wrap and runs to the terminal on the component shown on page 104.

Note: While you're making this first harness drawing, don't worry about where each wire goes after it enters the harness wrapping. Just make sure that you show the *exact number of wires* going from the harness wrap to each com-

ponent. For now, that's all that's necessary.

Incidentally, if you've made a super mess of your drawing, don't get upset. Set the project aside for awhile, then try again, using a clean sheet of paper.

If you've been working along as you read, you should now have some idea what I meant about patience. Hang in there. You're getting close to the end of the complicated part. There's just one more important step.

Get out the circuit drawings you completed in the last chapter, and make

Individual wires remain in main harness until they get close to component they will attach to. On your harness drawing, you'll indicate only the ends of the wires where they come out of the harness.

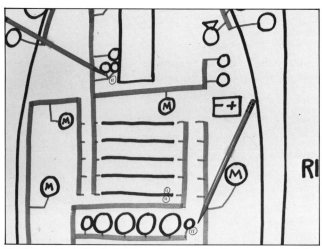

I've begun transferring information from basic power circuit diagram onto harness design drawing. I've used same number at both ends of the wire, where it comes out of the harness. Be careful not to use the same number twice, representing two different circuits; errors like that can be disastrous!

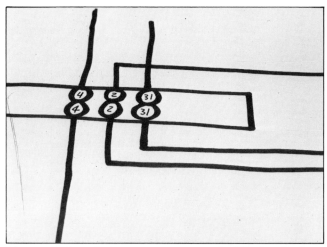

When a wire from the harness gets to a terminal block, write down wire's number on the block. Where circuit continues into second harness, use the *same* number to indicate that this is a continuation of same circuit. This means that one number represents the circuit, regardless of how many terminal blocks it passes through.

several photocopies of each. You're going to be writing on the copies, so keep the originals separate.

Start with a copy of your basic power-circuit drawing. Select any wire—individual circuit—on the drawing and find both ends of it. Number both ends with a 1. These numbers are for your use, so it doesn't matter which wire you start with. I like to put a circle around each number. Select another wire and number both ends 2. Continue until you've numbered both ends of each wire on the drawing.

Do not number any phantom (dashed) wires on the drawing. Each of these are numbered in the circuit drawing where they are drawn as solid lines.

If two different wires are connected to the same terminal of a component, use a separate number for each.

I find it convenient to keep a log sheet to indicate the name of the circuit associated with each number. This does two things. It helps me keep from using the same number twice, and the log can be used for reference later.

Repeat the wire-numbering process for each of your circuit diagrams, but do *not* start back at number 1 for each dia-

gram. Each wire in the electrical system should have its own number. It is not uncommon to use 50 or more numbers for a given car, so don't worry about how many numbers you're using. Just keep accurate records.

Here's the final tricky point to understand: So far, you've drawn each wire from one component to another in a single unbroken strand in the individual circuit diagram. But, because you'll probably be using the basic four or five harnesses joined together with terminal blocks, you must indicate terminal-block connections on the circuit drawings.

What this means is a circuit that begins in one harness could run through a terminal block and end up in another harness, even though the whole circuit is shown on one circuit drawing. For instance, your circuit for the oil-pressure sender will enter the engine harness, go through a terminal block on the firewall, then through the instrument-cluster harness to the oil-pressure gauge or indicator lamp.

When a wire in one harness gets to the terminal block, it stops. Number this end of the wire with the same number you used for the beginning of the wire. Use

the same number *again* where the circuit continues at the other side of the terminal block and goes into a different harness. Continue using the same number for that circuit until you reach the other end, whether it goes through a terminal block or from one harness to another.

Now, transfer the wire numbers from the basic-power-circuit drawing, one at a time, onto the harness-layout drawing. Remember to indicate wire numbers at each terminal block as well. See photo above right. Clearly mark each wire on your circuit drawing as you transfer the numbers to the harness layout. I use a yellow felt-tip marker for this.

This transferring step will take quite a while, so be prepared to take your time. If at all possible, have someone check your work, circuit by circuit, to eliminate errors. When you've finished numbering the harness-layout drawing, it will be your final authority for wiring your car, so it must be 100% accurate.

When you *think* you're done, check over each circuit drawing to be sure each wire is colored in, indicating that you transferred it to the harness layout. Also, check each dotted wire to make sure it is shown solid, and colored, on one of the other drawings.

After you've triple-checked everything, celebrate! The hardest part is done. All you must do now is "connect the dots."

CONSTRUCTING THE HARNESS

Your harness-layout drawing should now be complete and double-checked for accuracy. From here on out, it will be your guide in completing the electrical system. This is the part I refer to as *connecting the dots.*

There are two ways to install a harness: You can assemble the harnesses outside the car and install each as a unit, or you can install each wire directly into the car, then bundle the wires into harnesses. Both methods are described in this chapter, and there are arguments for and against each. I'll present both sides so you can decide which method is best for you.

Making a Harness Outside the Car— This method involves making a *harness fixture* from pegs or nails attached to a plywood sheet or similar material. The fixture is used as a guide to construct the harness. If you're going to wire several similar cars, or you and a friend are building similar cars, the out-of-car method may be preferable because the same fixture can be used for constructing more than one harness. However, if the wiring job is relatively simple, this method may be more work than it's worth. Installing wires one-at-a-time makes better sense. The latter method is also the way to install accessories to an existing system. Other than these considerations, here are the advantages and disadvantages to the out-of-car method.

The good points are:
● Creating subassemblies allows you to carry an individual task to conclusion before going on to another.
● It's much easier to string wires on a flat board—the harness fixture—than through a car.
● Generally, it's easier to get an overview of what you're doing, so errors are easier to spot.
● You don't have to continually change tasks. That is, first you'll measure and

cut all the wires for the harness. Then you will install all the terminals and shrink wrap, and so on. By keeping the tasks simple and discrete, chances for error are reduced.

The bad points are:
● You must label each wire with masking tape or other label as you go along. If labels get removed accidentally, you could end up with a mess.
● You can't cut the wires to exact length until the harness is in the car, so there will be some waste.
● If you take inaccurate preliminary measurements, you could end up with a *lot* of waste or short wires.

Installing Wires Individually—The good points are:
● Depending on your working style, you may find that you can work more accurately if you install and connect each wire to the appropriate component at both ends as you go, then move on to another wire.
● Because wires don't need to be labeled, you needn't worry about labels getting knocked off and the resulting confusion this can cause.
● Because wires are installed directly from one terminal to the next, you won't have to plan for excess wire, resulting in minimal waste.

The bad points are:
● You'll be repeating the same motions over and over again as you route each wire separately. You'll also have to constantly switch tasks—from cutting and stripping wire to terminal crimping and shrink-wrapping, to connecting terminals. This may or may not be a disadvantage to you.
● You must check and double-check each wire as you install it. After its terminal is attached and connected, it's too late to make changes easily.
● Harness wraps are harder to install. If you use loom tape as your final wrap, it can be almost impossible to install in tight areas. Also, it's difficult to install flexible conduit to wires after they're in the car, especially if you're trying to install conduit to a whole bundle of wires at once, in a tight area.
● To prevent unnecessary confusion, you may need to use temporary ties to keep the wires neatly bundled as you go.

OUT-OF-CAR HARNESS

The first step is to make a clothesline mock-up of each individual harness. Use plastic-covered clothesline. Start at one of the terminal-block locations and route the clothesline through the same path the finished harness will follow. Where branches lead off from the main harness, tape additional lengths of clothesline as needed. Install the clothesline mock-up in the car with nylon ties, clips or whatever you'll use to hold the finished harness in place. This ensures accuracy of length and location of the various branches.

Locate each place where a wire, or set of wires, will lead from the harness to a specific component. Mark that spot on the clothesline with an indelible marking pen. Attach a masking-tape flag next to the mark. On the flag, write the name of the component and the distance in inches from the branch point to the component. Take precise measurements. Later, when you make the harness, you'll allow extra wire for adjustments.

The clothesline-and-marker mock-up will closely resemble the harness layout you made earlier. As you mark each component's branch point on the clothesline, use a highlighter pen to check off that point on the harness-layout drawing.

When you're done making the clothesline mock-up of the first harness, carefully remove it from the car and put it in a safe place. Start at the other side of the terminal block and repeat this process

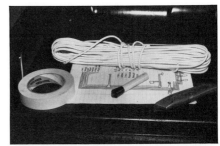

These are the tools you'll need to "wire" your car with clothesline. I prefer plastic-covered clothesline. Use a utility knife to cut it. Masking or duct tape can be used to attach branch points to main harness, as well as to attach "harness" to car. You can also write notes on the tape and attach them to the clothesline. Use a waterproof marker to write your notes. I also keep my circuit schematics handy for reference.

Here's an under-dash clothesline mock-up going into place.

Use masking tape to hold the clothesline in place. If possible, try to represent actual clamps or nylon ties that will hold finished harness in place.

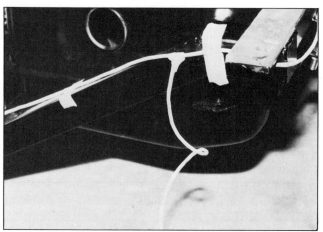
Notice how branch point has been taped to main line of harness. This represents location of taillamp branch in finished harness.

The clothesline is installed to represent final harness routing. Note how line is run inside chassis rail for added protection. Branch point leads to fuel-gauge sender. It must not interfere with movement of rear axle.

until you've mocked up each harness.

Making a Harness Fixture—In the next step you're going to fabricate a fixture used to construct the harness. The materials you use for the fixture are up to you. Gypsum wallboard (drywall), plywood, heavy cardboard and pegboard all work well. I think golf tees and garbage-bag ties on pegboard work about the best. The size of the fixture can be determined by stretching out your largest clothesline mock-up and measuring it.

Position the first of the clothesline mock-ups on the board. Stretch it out straight, so you can position the branches at the correct angle to the main trunk.

Temporarily fasten the mock-up in place. Use a generous number of tees or nails to make channels for the finished harness to be formed within. Use your marker to indicate all branch points and end points of the harness on the board, then remove the mock-up and keep it for future reference.

Return to the harness-layout drawing. Identify the first wire to be installed. Select the appropriate gage—and color, if you're going to color-code your wires. Determine where on the board the wire should start and temporarily fasten the end of the wire. *Allow 3 inches of excess* to extend past the starting point you

marked earlier. Follow the path established by the mock-up until you reach the point where the wire should stop. Cut the wire, again leaving 3 inches of excess. Use tags to label both ends of the wire with the appropriate circuit number, and component name, if you want. Then, indicate on the layout drawing that you've installed that wire. Repeat this process until you've installed and labeled each wire in the harness. NOTE: *Do not* install the terminals on the wires, yet. This is done after the harness is in the car.

Decide where *grommets* will be needed on each of the harnesses. Use a grommet wherever a harness passes

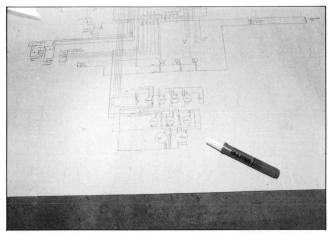

Refer to circuit schematic often as you begin installing wires into your harness.

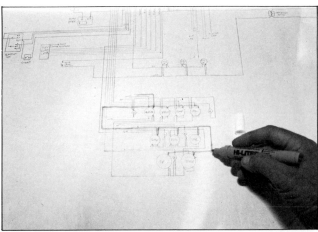

Use a marker to indicate on circuit diagram that an individual wire has been installed into harness.

through a hole in the body or frame. It will prevent the wires from chafing on the edges of the hole. Also, it bundles the wires tightly to minimize air or water leakage through the hole.

The correct-size grommet is one that's barely big enough for all the wires to pass through. And the groove around the outside edge of the grommet is the correct width and depth to seal against the edges of the hole. For more on selecting grommets, see page 66.

When you're sliding grommets onto the wire bundles, you may find that the tags you've put at the ends of the wires prevent the wires from feeding through the grommet. If this is the case, *temporarily* remove one tag, feed the wire, then replace the tag before going to another wire.

At this point, take a break. Then, double-check each wire, comparing it with the circuit drawing, harness layout and clothesline mock-up. When you're satisfied that the wiring is correct, bundle the wires together with nylon ties, electrical conduit or other harness wrap of your choice.

If using ties, strap each branch point just tight enough to prevent unnecessary movement, but loose enough to permit final adjustment when the harness is installed in the car. If you're using loom tape as a final wrap, overlap each wrap by at least 1/3 the width of the tape. But be forewarned: It's next to impossible to make final adjustments to a harness that's

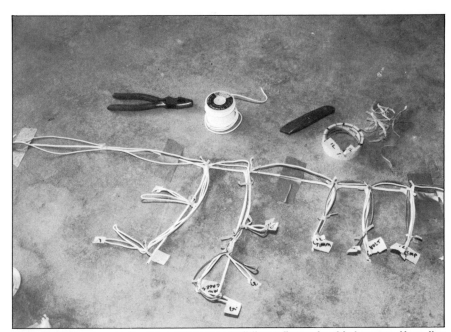

I taped my mock-up to the floor rather than building a fixture for this harness. Here, I've begun installing wiring over clothesline. During this phase, I use garbage bag ties to keep harness bundled.

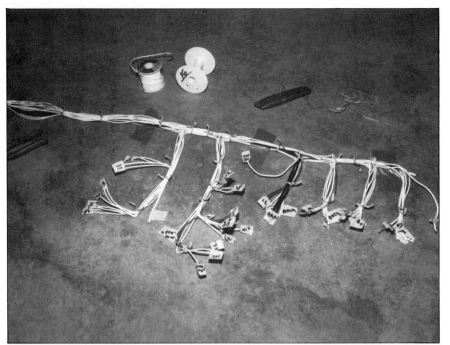

More wires have been added. Paper-tape tags on wire ends indicate name of component wires will attach to. You may prefer to write circuit number instead of component name—in any case, label all wires.

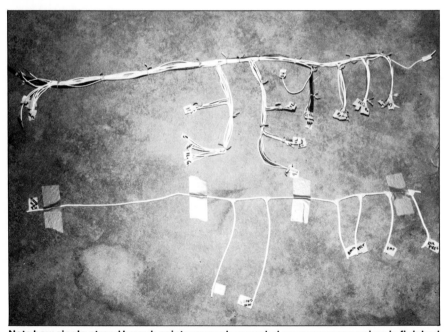

Note how single-strand branch points on mock-up ended up as numerous wires in finished harness. This is my instrument-panel switch harness. Vertical portions in center will attach to screws on various terminal blocks inside console. Left end attaches to headlamp and wiper controls, while branches at right are used to control interior lighting, heater blower and other accessories.

taped. Frankly, I prefer the nylon ties and strongly recommend them.

Repeat the bundling process until all your harnesses are assembled.

Before continuing, let's pause to make several observations. As mentioned earlier, if you're wiring several similar cars, or you and several friends are wiring similar cars, there are advantages to making a fixture and assembling the harnesses outside the car. Essentially, you'll have a small factory-type assembly line going. Be warned, however, that you should install the first harness in one of the cars to make sure it fits, then make any necessary adjustments to the fixture before duplicating the rest of the harnesses.

After all your harnesses are assembled, turn to the next chapter and read about how to install them in the car and to test them for correct operation.

IN-CAR INSTALLATION

This section gives an overview of how to install individual wires into the car, one at a time. If you've decided to use this method, read the following chapter before continuing. Although that chapter deals largely with the out-of-car procedure, many of the installation tips apply to in-car installation.

The procedure works like this: Referring to your harness-layout drawing, determine where each harness will pass through the body. Use the harness-layout drawing to help determine this.

Drill a hole of the appropriate size and install a grommet to prevent chafing. If you're not positive what size grommet to use, count the number of wires that will pass through the grommet, using your harness layout as a guide. Then grasp a handful of wire scraps equal in number and diameter to the wires that will pass through the grommet. Measure the diameter of the bundled wire scraps and select the appropriate grommet.

If you guess wrong, you won't know until it's too late. If the grommet is too large, wedge some electrical tape or silicone sealer into the grommet opening to take up the extra space. If the grommet is too small, you must either add another grommet next to the first one to route the remaining wires through, or remove all

Install grommets on wires where they will pass through body. Be sure grommet is just large enough for wires, but small enough to minimize intrusion of water and wind.

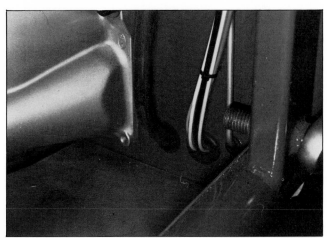

Grommets are used here to protect wiring harness and large battery cable where they pass through sheet metal. Because harness and cable together would have been a large bundle, we routed them through separate grommets.

the wires you've installed and redrill for a larger grommet.

Refer to your harness-layout drawing and select one circuit. Locate the component on the car at one end of the circuit and attach a wire, using the appropriate terminal and shrink wrap, if you're using it. Route the wire to wherever it ends and install another terminal. Connect the wire to the component. The wire is now installed. Check off the wire on the harness drawing. Repeat this procedure until each wire has been installed and checked off.

You'll probably be tempted now to install the appropriate harness ties, tape or flexible conduit. Resist doing that. Leave the covering off until you've had a chance to test the circuits, in case you must change something. Test procedures are covered in the next chapter.

When testing is complete, install the harness wrapping of your choice. Installing nylon ties every 6 to 12 inches is probably the easiest method of bundling wires once they're in the car. Loom tape or flexible conduit is more difficult to install, but use one of these wrappings if you can install it.

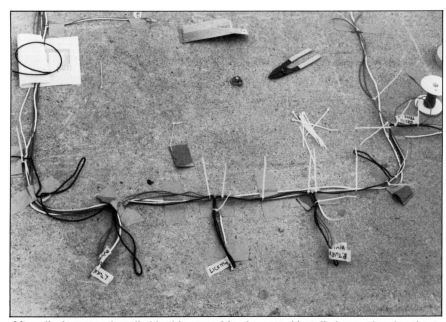

After all wires were installed in this rear wiring harness, I installed enough nylon ties to maintain its shape while being installed into car.

Installing a Harness

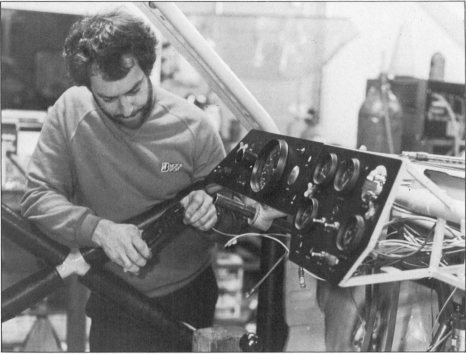

Multi-use tool cuts wire, strips insulation and crimps solderless terminals. The procedure is described on pages 115-117. Photo by Tom Monroe.

This chapter tells you how to install wiring harnesses. If you've decided to install the wires directly into the car, one wire at a time, the general procedures in much of this chapter don't apply. But, read this chapter anyway, because there are many installation tips that do apply, such as attaching the harness to the car, installing wire terminals and testing circuits.

The installation of a wiring harness follows this general procedure:
1. Rough installation of harness.
2. Permanent harness attachment.
3. Terminal installation.
4. Harness testing.
5. Making the connections.
6. Final testing.

PRELIMINARY STEPS

Check the holes where harnesses are going to pass through. Each should be the correct diameter to seal into the bottom of the grommet's groove. Also, smooth any rough edges to ease grommet installation.

Recheck each harness to be sure every wire end has a tag to identify where it is supposed to attach. If any tags are missing, replace them now. Don't wait until the harness is in the car.

If your work area and the car's interior are cluttered with tools and refuse, clean them up before continuing. You'll find that harness installation goes much better if there is nothing in your way. Also, gather up all the tools and materials

you'll need to install the harness and put them where they're handy. Nothing can wreck this part of the job any faster than discovering partway through that you're missing important items.

ROUGH INSTALLATION

By now you should know exactly where each of the harnesses will go. If not, this is the time to check.

Select one harness and put it into rough position in the car. For example, if you've picked the front-lighting harness, it will probably run across the front of the car, then rearward along one edge of the engine compartment, or trunk, if the car is rear-engine. Then it will pass through the bulkhead and probably end at a ter-

110

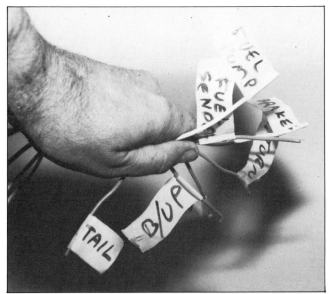

The first step is to be sure all wires are labled. It's far easier to check this out before you install the harness.

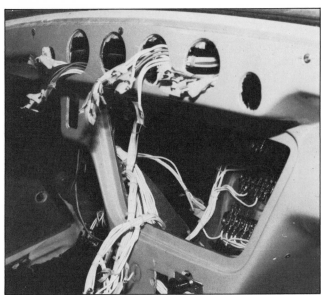

Put harness into "rough" position. When you're satisfied that harness will reach all its components properly, begin making hookups.

This harness is taped into rough position. When fit has been checked, install support clamps or ties.

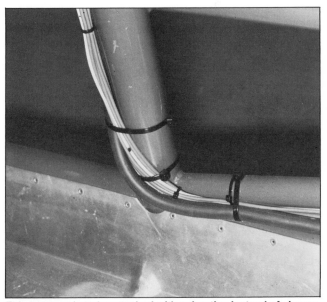

This harness is to be attached with nylon ties instead of clamps. Here, the first few ties have been installed to support harness in final position.

minal block or fuse panel somewhere under the instrument panel. It is evident that you should put the main part of the harness *outside* the passenger compartment and pull the short end into it. It wouldn't make much sense to put the harness inside the passenger compartment, then pull most of it through the grommet hole.

Install some clips or ties to secure the harness to the frame and body, but don't tighten these yet. Leave the assembly loose enough for final adjustments. Don't worry about the wire ends, other than being sure that all the identifying tags are still attached. *Do not* install any terminals now. Each wire should comfortably reach its component, with about

3 inches of excess—which you should have provided for when assembling the harness on your assembly board earlier.

Repeat this procedure with each harness that passes through the body. These will probably be the engine-compartment harness and rear-lighting harness. Position the instrument-panel harness last. In the case of my car, this was a

I prefer rivets to screws when installing harness clamps to body or chassis. Rivets and rivet guns are available at most hardware stores. For economy, purchase rivets in boxes of 50 or 100.

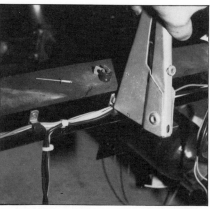

To operate gun, insert rivet and squeeze. Rivet fits into pre-drilled hole in chassis. When correctly installed, rivet is superior to screw because it can't vibrate loose. I find that it's easier to squeeze a rivet gun than it is to repeatedly turn a screwdriver or wrench, too.

Tie mounts are available in both self-adhering and screw-on types. Both work the same way—attach tie mount to car, then thread nylon tie through it and around harness. I used self-adhering type here, because harness is being attached to outer aluminum skin of this tail section.

two-part assembly, with one harness for the gauges and one for the switches.

Harness-Attachment Devices—Several types of devices are available for attaching harnesses to the car's body or chassis. The type(s) you use for each location is determined by the requirements for that application.

But, regardless of the attaching method, the device should not be so sharp that it cuts through the wire insulation. It should also be wide enough to spread the clamping load over a large enough area of the wires to prevent distorting the insulation. For these reasons, don't use bailing wire or thin wire clips to retain wires and harnesses in place.

Attachment devices include nylon ties, wiring straps, rubber-insulated metal clamps, nylon cable clamps and push-on frame clips. These are described on page 69.

If you're using nylon ties to bundle the harness, you can also use them to attach the harness to the car body. One example is to secure a harness at an edge or hole in a panel it runs through.

If the harness doesn't run close to an edge, you'll need to drill two holes in the panel to install this type of tie. Feed the tie into one hole, run it behind the panel, and out the other hole. Then, wrap the ends around the harness.

Although this is a serviceable attaching method, it tends to look tacky—

other attachment devices give a more finished appearance. Or, you can use nylon ties in conjunction with *tie mounts*.

Tie mounts provide a way to attach a nylon tie to a surface where you don't want to drill two holes, or one where you cannot get behind the panel to feed the nylon tie back through. For this reason, tie mounts are a good compromise—better than a nylon tie alone, but not as good as a clamp.

The plastic tie mount attaches to the car with a rivet or screw. The nylon tie is looped through slots in the mount. Some types are self-adhering, but I don't recommend these. Tie mounts are available to accommodate one or several ties on one mount. An installation using both tie mounts and nylon ties looks fairly "finished," while permitting easy adjustment and replacement, if necessary.

One positive thing about nylon ties is they can be left loose until you're ready for final tightening. And, if you get one too tight, too soon, they are easy to cut off and replace. Some nylon ties have an enlarged end with a mounting hole. These can be installed with a rivet or screw.

Custom fittings are available for some

Here you can see extensive use of nylon ties, both to bundle harness and support it. Note how large ties permitted attaching harness bundle to gauge clamps.

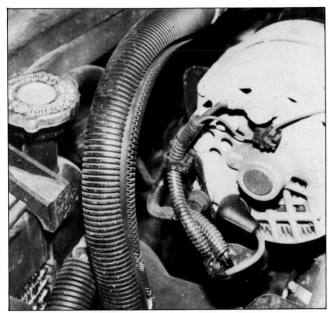

OEM conduit is installed with custom fittings. Some of these are available in aftermarket.

Taillamp wiring is temporarily retained by tape until suitable fasteners can be installed. Never use tape as a permanent fastener.

brands of flexible conduit. Such fittings are intended to be used with specific conduits of a given diameter, and at specific junctions of the conduit. If you can find them for the type of conduit you're using, they are ideal.

To install one of these fittings, first drill a mounting hole in the panel. Then attach the fitting to the conduit and push the expansion lug into the hole. A barb on the lug pops out against the backside of the hole to retain the fitting. If you try to remove one of these fittings, you may destroy the lug when prying it out, so they're usually not reusable.

Tape of any kind is a poor method of attaching a harness or wire. Don't use it! But, electrical, masking or "200-mph" tape is useful to hold things in place *temporarily* until you're ready to make final adjustments and install the appropriate attachment device.

PERMANENT ATTACHMENT

Once you have *all* harnesses loosely installed, inspect each to see that the branch points permit the wires to leave the main harness smoothly. Make sure each branch point is situated to permit the wires to run to the components without strain, unnecessary slack or excess wire

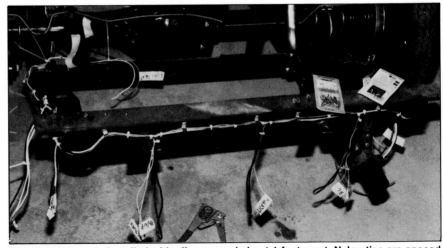

Chassis harness is installed with clips spaced about 1 foot apart. Nylon ties are spaced about 4 inches apart. Both dimensions are about ideal.

length. Also, make sure the main harnesses and the branches are satisfactorily routed away from areas of excess heat, sharp edges, moving parts and other mechanical factors that could damage the harness.

Bundling the Harness—Now, carefully add more nylon ties to each wire bundle to achieve a neat appearance. This may require a tie every 6 inches or less. Work carefully. Check frequently that

you haven't created any unsightly bulges and haven't added or decreased slack in any of the branches. Add ties until you reach a distance of about 12 inches from each component.

In most cases, it's best to install all the additional ties fairly loosely, then go back and tighten them gradually, in sequence. Continue this process until all ties are tight. However, don't get them so tight that they distort the wire insulation.

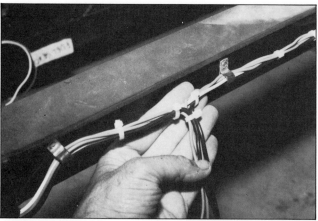

Cut off nylon tails with a wire cutter after snugging ties. This photo shows correct method of installing ties at a branch point.

Fuel-sender harness is retained by a clamp under one of the fuel tank mounting bolts. Another clamp will be installed in other bolt hole to retain ground wire.

Loop of instrument panel harness will have to be supported to prevent excess vibration when car is being driven. Note how nylon ties have been used to keep harness in neat bundle.

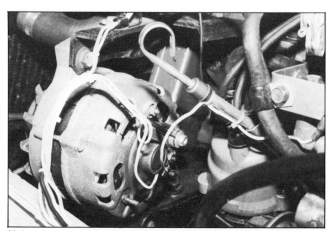

Nylon ties are used to retain electric choke feed wire to oil dipstick tube. It's better to use any available support rather than fabricating additional parts. Note slight amount of slack in generator wires to permit adjustment of drive belt.

Finally, use a diagonal wire cutter to cut the excess tails from all the ties. Cut off the tails flush with the clamping blocks of the ties. This looks neater and reduces the possibility of catching something, like your skin, on one of the ends later. I'm particularly thinking of how badly I cut my hands on one of my first harnesses. I wasn't careful, and the sharp little ends seemed to be everywhere.

Securing the Harness—Check each harness to see where additional support is needed. It's best to support the harnesses every 12 inches or so, using one of the harness-attachment devices discussed earlier. As you install extra supports, make sure you don't accidentally pull the harness out of place or put unnecessary strain on any wires.

Once all harnesses are permanently installed, the electrical system should look about finished. And it is! There's only one more major task to do before you can test the system—install the terminals. Please note that *terminal installation* refers to attaching the terminals to the wire ends only. You won't be connecting the terminals to the components until *after* you've tested the harness.

Although the processes of terminal installation and harness testing and connecting are described as three separate procedures, you may find it more convenient to combine all three procedures on each wire or each circuit as you go. If you work carefully, this method is perfectly OK. But, carefully read each of the following procedures before deciding whether or not to combine them.

TERMINAL INSTALLATION

Before you start installing terminals, have all the required tools and materials within easy reach. Some components will require specially shaped connectors. These components include the generator, voltage regulator, ignition switch and turn-signal switch. If the required connectors are not readily available at an auto-parts store, you can cut them from the wiring harness of a junked car. Be sure to leave at least 6 inches of wire attached to the connector. The terminals on other components in the system usually use the standard connectors described on pages 63-64.

Generally, terminals come in two

When possible, use custom terminals. This headlamp terminal is a good example. I could have used three female terminals just as well, but this one eliminates possiblilty of making wrong connections.

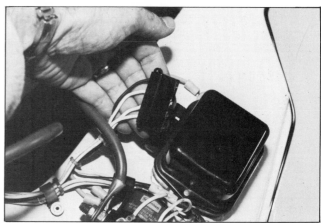

Here's another example of a custom terminal. In this case, individual terminals would not have been satisfactory. Should voltage regulator need replacement, custom terminal is easy to connect. It can go in only one way.

types—*solderless* and *soldered*. To help you decide which to use, the good and bad points of each type are listed on the next page. If you think the information is biased in favor of solderless terminals, you're right. I prefer them, and so do most people who've done much wiring. But the final choice is yours.

Installing Solderless Terminals— Correct installation of a solderless terminal requires two tools: one to strip the wire insulation and another to squeeze the terminal barrel. If you're going to do just a few terminals, you can strip the insulation with a knife and do the crimping with a pair of diagonal wire cutters. But I don't recommend either. The correct tools for the job are relatively inexpensive and readily available at most auto-parts stores, hardware stores and electrical-supply houses.

Wire strippers are available in several forms. The least expensive type looks like a pair of pliers with unusual-looking jaws. When the jaws are closed, it looks like a series of holes have been drilled through the line where the jaws meet. The holes are sized according to wire gage. These are often built into the crimping tool.

If you've never used wire strippers, practice on some scrap wire. With the jaws open, put the piece of insulated wire in the correct opening. Close the jaws and pull. If you selected the right opening, the jaws should cut and remove the insulation without harming the wire inside. If not, try another size. When done correctly, the stripping operation is quick and feels right.

Another type of stripper has a jaw that adjusts for each wire size. These are excellent tools. Many experienced electricians prefer them, having one preset tool for each wire size.

A more expensive wire stripper is essentially automatic. You insert the wire end—any size within reason—into the tool and squeeze. It grasps the wire, sizes it, and cuts and discards the insulation,

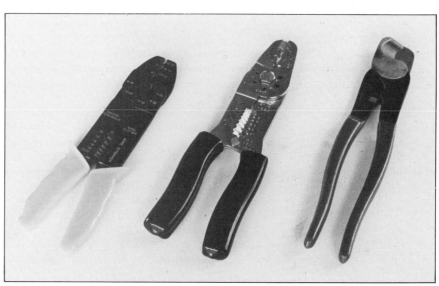

Tools required for installing terminals include a wire cutter, stripper and crimper. Two tools on left do all three. Photo by Ron Sessions.

Here's a hot tip for keeping your supply of terminals neat while installing them on harness. Steal (OK, borrow) your kid's frisbee and use it as a dish. It's shallow enough not to tip over easily, it's unbreakable, and you can put a handful of several types of terminals in it at once without them getting mixed up.

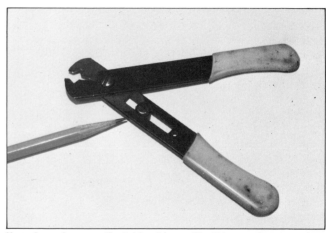

If you're using an adjustable stripper like this one, loosen screw and adjust jaws to obtain good strip without damaging wire strands. Then tighten it. Only drawback is that stripper is good for only one wire size at a time.

all in less time than it takes to describe it. This kind of tool is nice if you have a lot of work to do, but it's more expensive—at least $10.

Terminal crimpers are nearly all alike, but as with any handtool, price varies with quality. Choose a crimper with a selection of die sizes that will accommodate all the terminals you'll be using. Make sure it can be squeezed and opened smoothly. It should have rubber handle grips for extra comfort. Most crimpers also have a cutter blade, so you don't have to buy and keep track of a separate wire cutter. Some terminal crimpers also include a wire stripper as part of the tool design. You may get lucky and find one that works well—many don't.

Solderless terminals come in an incredible variety of sizes, shapes and types. Fortunately, you'll only need a limited assortment for wiring your car. Those used for automotive wiring are discussed on pages 63-64.

When selecting terminals, first consider the *barrel size*. The barrel is the tubular part of the terminal that the wire fits into. Obviously, the wire should fit snugly in the barrel. Generally, terminals are made to accommodate at least two wire sizes: 16—14 gage, for example. The number before the hyphen is the smallest wire size the terminal will fit; the number after the hyphen is the largest. Remember that wire gages are ex-

pressed in even numbers only.

Second, consider whether you'll use terminals with insulated or non-insulated barrels. A non-insulated barrel simply slips onto the wire and is crimped. To insulate this type of terminal, add a piece of *shrink wrap*. Insulated terminals have a built-in piece of insulation for the wire to slip into. The inside diameter (ID) of the barrel insulation allows the wire insulation to fit inside. And, the insulation continues for the entire length of the barrel. In the case of quick-disconnect insulated terminals, the female terminal is insulated over its entire length.

Shrink wrap is a tubular plastic material that shrinks when heat or flame is applied to it. It is not harmed by direct flame from a match or cigarette lighter. In terminal installation, shrink wrap is used to provide both insulation and extra strength to the connection.

Shrink wrap can be installed over insulated terminals, but it's a waste of money to buy such terminals if you're using shrink wrap. However, if you are using insulated terminals, shrink wrap will strengthen the connection.

Cut a piece shrink wrap to length and slide it on the wire *before* installing the terminal. After crimping the terminal, slide the shrink wrap onto the terminal and briefly apply heat or flame—the wrap shrinks in a moment. A cigarette lighter is perfect for this job.

You might wonder why anyone would

COMPARISON OF SOLDERLESS AND SOLDERED TERMINALS

Solderless (Crimp) Terminals
Good Points
- Can be installed immediately, without warm-up time.
- Can be installed with inexpensive crimping tool.
- Are widely available in all standard and many custom types.
- Are used in nearly all factory installations.
- Require little manual skill to install correctly.
- High-quality types offer excellent, permanent connection.

Bad Points
- Difficult to remove without shortening wire.
- Can be installed wrong if inattentive.

Soldered Terminals
Good Points:
- Fairly easy to remove if installed wrong.
- Provide good electrical connection.

Bad Points:
- Require special equipment (soldering iron).
- Require special materials (solder and flux).
- Require skill and practice to do correctly.
- You can burn yourself or insulation, if careless.
- Can cause stress points in wires.

use shrink wrap when insulated terminals are available. The reason is that shrink wrap adds support to the joint between the wire and terminal. Its effect is similar to the spring or thick rubber covering used where an electric cord attaches to an appliance. It prevents the wire from bending too sharply at the connection. An added advantage of shrink wrap is it helps weatherproof the connection. In fact, the most expensive shrink wrap includes a coating of hot-melt adhesive inside for this purpose.

The Procedure—Select the right size and type of terminal—it should mate with the terminal or stud to which it attaches. If you're using shrink wrap, slip a piece onto the wire. Check the wire length and cut off any excess, but be sure to allow enough slack in the wire so it won't pull the terminals off the posts.

Next, strip the wire the amount required for the terminal you're using. If you're not sure how far back to strip the wire, hold it alongside the terminal barrel and align the barrel end with your thumb. Then strip the wire from that point to the end. Slip the terminal onto the stripped end and crimp the barrel with your crimping tool. Then slide the shrink wrap in place and heat it. You're done.

There's a right way and a wrong way to use a crimping tool. Here's the right way: When you place a terminal into the crimping tool, choose the die set that is closest in size to the barrel you're going to crimp. Also, the terminal barrel will have a slit or seam running its length.

On aircraft-quality terminals, and other terminals with high-quality barrels, the seam is silver-soldered or welded shut, so it is nearly invisible. In either case, be sure that the terminal is positioned so the *protruding die* contacts the barrel on the side *opposite* the seam. The reason for positioning the die like this so it doesn't punch into the seam, splitting it open, rather than flattening the barrel as it's designed to do.

Then, making sure the wire is fully inserted into the barrel, squeeze the dies together firmly and release. Tug lightly on the wire to make sure the crimp holds. It's only necessary to crimp one place, but if you decide to make two crimps, be sure the protruding die is positioned opposite the seam.

Installing Solderless Terminals

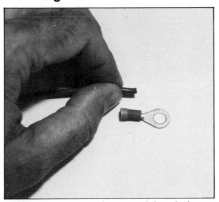

1. If you're not sure how much insulation to strip from wire, hold wire next to terminal and use thumb and finger to gauge length. Strip just enough so exposed end fills terminal barrel—no more, no less.

2. Strip the wire. Be sure you don't accidentally cut into wire strands themselves.

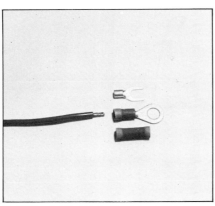

3. This wire is correctly stripped to fit into any of terminals shown.

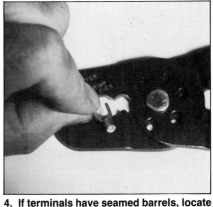

4. If terminals have seamed barrels, locate the seam. If terminal does not have a seam, locate where seam would be if it had one, or the side opposite where barrel flares out to form connector. Next step shows how to position in crimper.

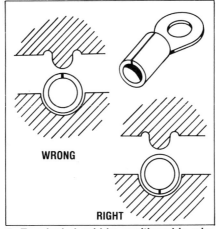

5. Terminal should be positioned in crimper so seam—or place seam would be—is *opposite* protruding die on crimper. The idea is to drive protruding die into side away from seam to avoid opening it.

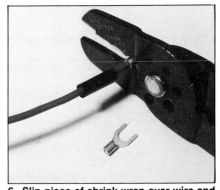

6. Slip piece of shrink wrap over wire and place terminal in crimper. Slide wire into terminal and crimp terminal onto wire using the appropriate die size in crimping tool. Most tools have at least two dies; some have three. *(continued next page)*

Continued from page 117

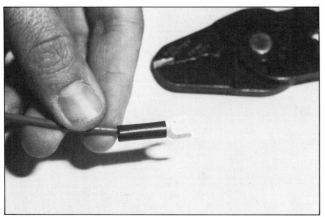

7. Slide shrink wrap into place and heat it with cigarette lighter, match or heat gun.

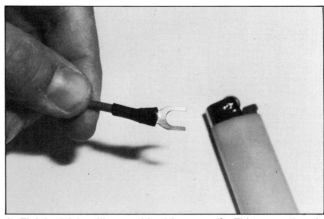

8. Finished job will resemble this example. This procedure is superior to a soldered joint.

Here, I've just crimped an insulated ring terminal.

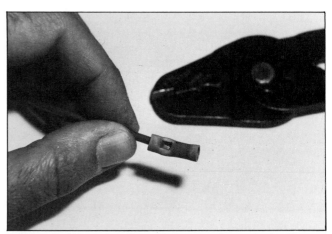

Butt terminal has to be crimped twice, once for each wire. Be sure to crimp only on narrow part in middle—enlarged ends are designed to fit over wire insulation.

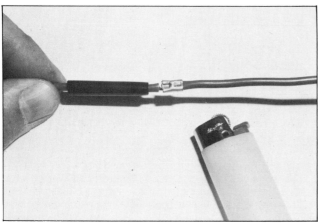

If you intend to shrink wrap a butt joint, use a non-insulated terminal.

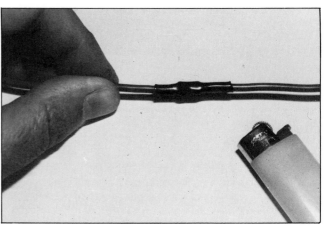

When shrink wrap is heated, it conforms tightly to butt terminal and wire insulation.

Soldering Terminals

Soldering wires requires a few tools—soldering iron (on glass ashtray), wire stripper, can of flux, piece of scrap wire for applying flux, spool of rosin-core solder—*do not* use acid-core solder intended for plumbing and tinsmithing—and disposable cigarette lighter for applying shrink wrap. I prefer a pencil-point soldering iron to a soldering gun.

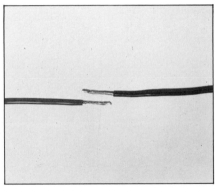

To join two wires, begin by stripping about 1 inch of insulation from both.

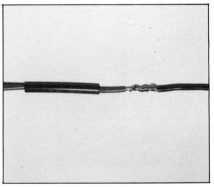

Slip piece of shrink wrap on one of the wires and twist exposed conductors together. The idea is to get a mechanical joint that would almost be strong enough to hold wires even without solder. This is *not* recommended, however!

Apply flux to wires. Flux acts as a cleanser that ensures good bond for solder. I use a piece of scrap wire to daub a bit of flux on the joint. Don't use too much.

Heat flux slightly with hot soldering iron until it flows into all the wire strands. Don't apply heat for too long, because flux will decompose, generating odor and smoke.

Heat joint with soldering iron. Note that I'm heating the joint, not the solder. When the joint is hot enough, I hold the solder against the joint and it melts into place.

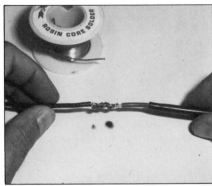

Tug on joint to be sure solder has held. Also look for evidence of a "cold" joint, which is a dull grey color. Reheat if necessary, until solder retains a slightly shiny appearance when cool.

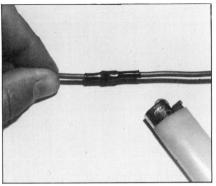

Heat shrink wrap with cigarette lighter, match or heat gun. I find that a disposable lighter is perfect for this. As soon as heat or open flame is applied, the wrap shrinks onto the joint.

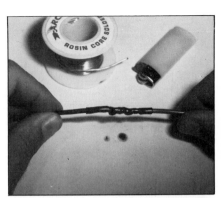

Finished job should look something like this. Shrink wrap is tight and extends onto wire insulation to prevent short circuits.

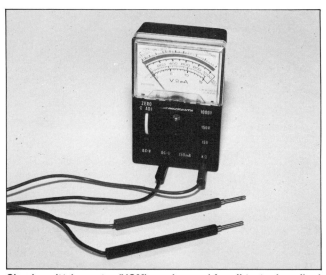

Simple volt/ohmmeter (VOM) can be used for all tests described here. I've used this inexpensive Radio Shack model on numerous, fairly sophisticated wiring jobs. It cost me less than $10. back around 1975.

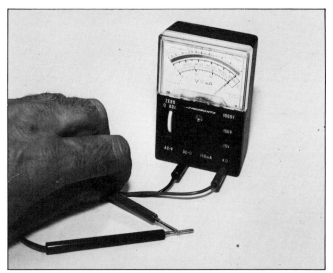

Always check calibration before using ohmmeter. Touch probes together, then turn calibration wheel until needle points to 0 ohms. This also checks the built-in battery.

TEST THE HARNESS

A beginning word of caution: *Do not hook up the battery yet!*

Before attaching any of the wires to the components, perform a test on each circuit in the harness to be sure there are no hidden problems. This task may seem tedious, but it's much easier to find and correct problems *before* you attach the wires to anything. No matter how careful you think you've been, there's always the possibility of error. At best, the component simply won't work. At worst, your new harness will go up in smoke. Enough said?

There are several methods used to test circuits, but each boils down to checking *wire continuity*. This simply means checking that the wire marked RIGHT FRONT TURN at one end is the same wire that's marked RIGHT FRONT TURN at the other. Be sure that none of the circuits got crossed during assembly or installation.

You can use a *self-powered test lamp* or an *ohmmeter* to test continuity. Both instruments tell you if the two wire ends are connected to each other. Before testing a wire for continuity, be sure the test instrument is working. Both the ohmmeter and the test lamp rely on a built-in battery for power. It's embarrassing to discover that the "faulty wire" you just

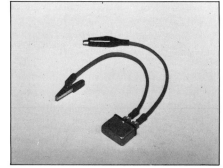

You can use a jumper wire like this to bypass controls when testing a circuit— never bypass a *load*. Note auto-reset circuit breaker I installed as a precaution. I sometimes use this wire to extend one of my ohmmeter leads when doing continuity checking.

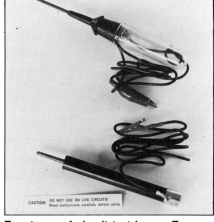

Two types of circuit test lamps. Top one indicates voltage in live circuits. Bottom one is battery-operated self-powered test lamp, used to check circuit continuity with power *off*. It's the one to use to check your harness as described in accompanying text. As package precaution says, *do not* use it on live circuits. Photo by Ron Sessions.

tested and replaced was OK, but the tester's battery was dead!

To check a self-powered test lamp, touch the alligator clip on the wire to the pointed probe on the end of the lamp body. If the bulb and battery are OK, the lamp will light. If it doesn't, determine why and fix it before proceeding. To check an ohmmeter, set the dial on OHMS. Some of the less-expensive ohmmeters use separate sockets for the

test wires for each of the meter's functions.

Now, touch the two test probes together. The meter needle should go to zero. If it does, the meter works. If the needle doesn't move, check that the wires are plugged in correctly and making good contact. If the connections are OK, the meter probably has a low or dead battery. If the needle moves, but doesn't go exactly to zero, calibrate the meter.

LED circuit tester is useful to detect voltage in live circuits. With lead grounded, LED lights if probe contacts voltage.

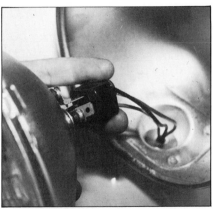

Connect components, one at a time, after verifying integrity of circuit.

MAKING CONNECTIONS

Pick any component as your starting point. Let's say you've chosen the right headlamp. If you've installed a special headlamp connector in the harness, it has three cavities that align with the three terminals on the back of the headlamp. Simply plug it onto the lamp and it's done. If you chose to use a separate connector for each wire, refer to the tags on the wires. Find the one for high beam and plug it onto the high-beam terminal of the bulb. Repeat this for the low beam and ground wires.

Note: Leave the tags in place until after you've tested each circuit with live voltage.

Essentially, the job of connecting the rest of the harness is a repeat of this procedure, over and over, until all components are connected. Refer to your circuit drawings as you work, just for the added comfort of an additional check. Mark off each wire as you connect it.

One place you must use extra caution is at terminal blocks where several harnesses join. Here, you'll have what seem like a hundred wires jammed into an impossibly small space.

To avoid confusion, clamp or clip each harness bundle into place at a right angle to the terminal block(s), allowing the separate wires to extend out randomly. Pull the entire handful of wires away from the terminal block and bend the wires a little, if necessary, to keep them away from the work area until you're ready to connect them. Then, find the first wire you want to connect, and shape it carefully and neatly to run to the terminal-block stud to which it attaches. Attach the wire to the stud. Repeat this process until all wires are attached. The wires should run in orderly rows between the terminal blocks. When you're done, carefully install additional nylon ties to these wires to keep them neat.

FINAL TESTING

The moment of truth has come! You've made visual checks of the harnesses, and performed continuity, ground and short tests of each wire. You've also connected each wire to its respective attaching point. What's left to do? Not much.

Install the fuses according to your cir-

There's either a screw or knurled knob for this purpose. Turn the adjustor until the needle lines up with the zero mark, then separate the test probes.

Select a wire in the harness. Locate both ends of the wire by referring to the tracers or tags you attached earlier. Touch one test probe to the terminal at each end of the wire. If the ohmmeter leads are too short, fabricate a jumper from a length of wire—14—18 gage is OK—and a pair of alligator clips soldered to the wire. The instrument should indicate continuity—lamp should light, or ohmmeter should indicate zero ohms. If the wire checks out, good. If not, inspect it to determine what's wrong.

If you get no reading, chances are you have tested one end of one wire and the other end of another. Move one of the probes to each of other nearby wires and you may find the one you're looking for. If this is the case, the wires may have been mislabeled. Or, there could be a serious error in routing. *Find the problem before continuing.*

Also, pay special attention to the terminals. Be sure the barrel of the terminal is crimped onto the *bare wire,* not the wire insulation. If it isn't, you'll get a faulty continuity reading.

After you're satisfied that you have verified both ends of one wire, perform a ground test on that wire. Leave one probe in contact with the wire. Move the other probe to the body or frame. Unless

you're testing a ground wire, there should be no continuity between the wire and ground. If there is, inspect all the metal clips the wire passes through to see if and where the insulation was damaged. You may be tempted to tape over the damage. Don't! Replace the entire wire.

Finally, test for shorts in the wire. To do this, leave the probe connected to one end of the wire. Move the probe at the other end to the ends of adjacent wires. There should be no continuity between wires. If there is, determine if the two wires are intentionally spliced somewhere in the harness. If not, look for and correct the short.

Also, look carefully at all places where the wires are clipped and tied. It's possible that one of the clips or ties tore holes in the insulation of two adjacent wires, allowing the conductors to short together. The only correction for this is to replace the defective wires.

Repeat the continuity, ground and short tests for each wire in each harness. Correct each problem as you find it. Otherwise, you'll be tracing problems after the circuits get complicated by the addition of all the switches, fuses, motors, gauges, lamps and such. And, although I've emphasized testing, it's possible that you won't find a single problem. If not, great!

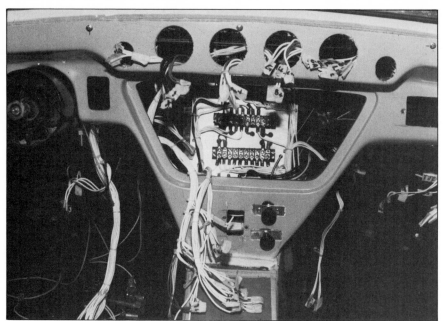

Wiring to this terminal block is kept neat by cutting wire lengths so they are just long enough to reach. If each wire were a random length, whole panel would be a mess. Note how wires run parallel to terminal blocks, then branch a right angle directly beneath terminal screw. It's fairly easy to trace circuits and locate troubles if installation is neat.

cuit diagrams. Then make sure all switches are turned off. Finally, attach the positive and negative cables to the battery.

At this point, nothing should happen. Look carefully and smell carefully. You should see no sparks or smoke, and should smell nothing burning. So far, so good.

Connect the negative probe of a voltmeter to the battery negative terminal. Or connect the pigtail of a 12-volt *non-self-powered* test lamp to the battery negative terminal. You will use this test setup to find out if battery voltage is available at the correct distribution points before you turn anything on. Again, use a jumper wire if the test leads are too short. Because this is a ground wire, you needn't worry if the alligator clip accidentally touches the ground. But keep it away from B+ circuits!

Refer to your drawing of the basic power-distribution circuit. It will show you the points where B+ should be available when the ignition switch is off. Generally, they are:

- Input stud of the starter solenoid
- Input to the ignition switch
- Input to the fuse-panel bus bar
- Input to the headlamp switch

Your car may have others. Probe each of these points with the test instrument. In each case, the lamp should light or the voltmeter should show 12 volts. If not, determine why and correct it. *Disconnect the battery before making any adjustments.*

Now, turn the ignition switch to the ON or RUN position. Probe each component that should get voltage in the RUN mode. These include the ignition coil, various accessory fuses, voltage regulator and voltmeter, to name the most common. Trace any discrepancies—again, disconnect the battery before adjusting anything.

With the ignition switch still in the RUN position, switch on each accessory to determine if it works correctly. If it does, turn off the accessory and go to the next one. If a fuse blows when you turn on any accessory, quickly turn it off and find the cause. It could be a short

circuit—although you should have found it during the harness test—or it could be a defective component. Also, make sure the fuse has the correct ampere rating.

Finally, make sure the transmission is in PARK or NEUTRAL, then turn the ignition switch to START. The engine should crank and start. With the engine running, check the instruments for correct operation. If you see a low oil-pressure reading, shut off the engine and don't restart it until you can correct the problem—there's no point in damaging a new engine. Check the operation of the charge-indicator lamp, ammeter and voltmeter. If you don't know how these should operate under normal conditions, see page 41 (charge indicator lamp) and 48 (ammeter and voltmeter).

There are two simple tasks left. Unless they are intended to be permanent, remove all the tags—you won't be needing them anymore. Put your circuit drawings and wiring-harness drawing in a safe place. They will come in handy if your car's electrical system ever needs service.

If you've gotten this far, the electrical system should be operating correctly. Be sure your vehicle is properly licensed and insured before doing a road test. After a few miles of driving, check your harnesses for chafing and correct the cause. Also, be on the lookout for places in the harness that need additional ties or clips, then install them. Check terminal connections for tightness and to be sure they aren't under any unnecessary strain.

Congratulations! You're now an official do-it-yourself electrical engineer! It may seem like it has been a long, hard struggle, but the end result is worth it. You have truly accomplished something that most people wouldn't think of trying. And the added advantage is that if anything goes wrong with your electrical system later, you'll know how to fix it—because you constructed the system in the first place.

Rewiring an Electrical System

Repro and custom harness kits are readily available in aftermarket for popular custom and collector's cars, like this 1966 Mustang GT powered by a High Performance 289. Other harness makers offer large selections of authentic repro harnesses for the more esoteric antiques and classics, or will make them to order. See supplier's list on pages 152-153. Photo by Tom Monroe.

Deciding to rewire a car is somewhat like deciding to take a trip. You must start with some basic determinations. For instance, where do you want to end up? How long do you want to spend getting there? How much time and money do you want to spend? Where can you get information on how to get there? Let's look at these one at a time.

First, where do you want to end up? Do you want to construct a like-original wiring installation? Or do you just need one that is serviceable? If you're restoring an expensive collector car, the answer is cut and dried—go original. But if the car is simply a tired but roadworthy "beater," you may decide to replace one or more wires without regard for wire color or insulation type, just so long as they work.

Second, how long do you want to spend getting there, and how much

money do you want to spend? In other words, how extensive a job do you intend to undertake?

A word of caution here: Rewiring jobs are never as simple as you first estimate. There's always something unforeseen to take up your time. And often, this turns out to be pride of workmanship. Once you get started, you may not want to stop until you've restored the entire electrical system! Of course, replacing a single wire that's chafed or cut is relatively easy. But if the whole harness was gnawed by mice or pack rats, who knows what may be involved?

Also, be sure you know *why* the job is needed. Are you restoring a car to 100-point condition? Are the wires dried out, chafed or oil soaked, but otherwise OK? Are you hoping to prevent an imminent burn-out? Or worse, are you too late, and the system has already gone up in smoke?

Finally, what kind of reference material is available? Can you obtain the manufacturer's shop manual? Or must you rely on an aftermarket manual from a library or bookstore? Can you get the information you need from a car-club member or obtain restorer's literature? Is there a similar car available for use as a model? Or, if all else fails, will you have to refer to the original harness and attempt to duplicate it?

In addition to the wiring, there's also the matter of restoring or rebuilding the components. All these considerations are discussed in the following sections.

TYPES OF WIRING REPAIRS

Generally, wiring repairs fall into one of three categories: *burn-out repair, selective wire replacement,* or *whole-harness replacement.* Each repair requires a different approach.

Here, you're looking at a complete rewire job—braided-cloth insulation is literally rotting off wires of front lighting harness. Terminals are likewise beyond help. Restorer of this "classic" will probably opt to replace original with an authentic repro harness. Photo by Jim Barrett.

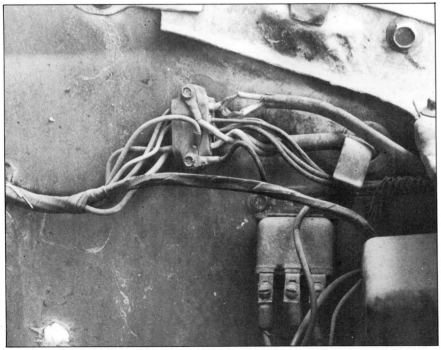

Most wiring problems happen at connections and wire ends. Corroded terminals are culprits of voltage drop; bare conductors can cause shorts. Cleaning or replacing corroded terminals and replacing components that are "about to go" should make this electrical system serviceable. Unless owner is restoring car to 100-point condition, wires inside harness are probably otherwise OK. Photo by Jim Barrett.

Burn-Out Repair—This type of repair can involve a single wire, a whole harness or anything in between. Often, when a single wire burns out, it also burns out all the wires wrapped in the harness with it. Regardless, the first step in replacing a burned-out wire is to determine why it burned out in the first place.

Wires usually burn out because they overheated. This often happens because the wire carried more current than it was rated for. If the load is a motor, the motor may have seized or slowed because of worn bearings or bushings, which can cause the motor to draw excess current.

Ideally, the fuse in the circuit should have blown before the wire burned out, but occasionally the unexplained occurs. In other cases, the amperage rating of the fuse is too high for the circuit it's supposed to protect.

If the load is not a motor, chances are there is a short to ground between the battery and load, probably between the switch and load. Such a short can be caused by chafed or worn-out insulation that exposes the bare conductor to a sharp metal edge or other ground. After the wire burns up and the insulation is gone, the cause may be difficult or impossible to pinpoint with certainty. In this case, protect the replacement wire over its entire length from anything even remotely threatening.

Carefully check for hidden damage. A wire that burns out got hot over its entire length. This can melt or burn the insulation on adjacent wires, even if hidden inside a harness covering.

Finally, look for evidence of other components or wires that are "about to go." If one component has aged enough to cause a burn-out, chances are, others have the same potential. Look for discolored or distorted insulation that could indicate an overheated wire that hasn't actually burned yet. Also check for corroded connectors.

Copper has a tendency to turn green or white when it oxidizes. Light green crumbs indicate a potential problem. This could be at a terminal, or the crumbs could be working their way out of a break in the wire's insulation. For information on how to check the system for bad terminals, see Diagnosing Voltage Drops, pages 129-130.

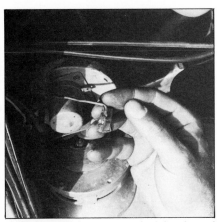

Melted insulation on terminal is sign of burnout. In such a case, don't take chances! Replace wire over entire length, find cause of overheating—often a load drawing too much current—and fix it. Photo by Jim Barrett.

Once you've found and corrected the cause of the burn-out, as well as other potential burn-outs, you're ready to replace the bad wire or wires. After you've determined the amount of work needed, refer either to the section below on replacing selected wires, or on replacing complete harnesses.

Replacing Selected Wires—Whether you're replacing a burned-out wire or one that's chafed, cut or otherwise damaged, the procedure is essentially the same.

First, if you're replacing more than one wire, remove and reinstall the wires *one at a time*. This will eliminate the possibility of creating new problems by miswiring something.

Second, it's best to replace the damaged wire over its entire length. This may entail unwrapping a harness, unless the covering is corrugated-plastic conduit. If the task is formidable, and you're not restoring a show car, you can cut off the defective wire at both ends so the ends don't project from the harness wrapping. Use additional tape, if necessary, to cover the cut ends. This has two purposes. One, it looks neater. Two, it prevents the old wire from accidentally contacting any source of current, which could cause additional problems.

Third, the replacement wire should be the same gage as the original, and if possible, have the same insulation color. Don't make the mistake of using a wire that's too small, or you may be making

another repair sooner than you'd planned! The color is of secondary importance. The wrong color could cause future problems only if you or some future owner tries to trace the circuit in question.

Finally, install the replacement wire into the original harness, using the appropriate terminal at each end. Replace the harness wrap. If you left the original wire inside the harness, tape the replacement wire to the outside of the harness instead of trying to insert it inside the original wrap.

Replacing an Entire Harness—To replace a harness, either construct the new harness yourself or buy a replica harness from a specialty supplier. Some suppliers offer exact replica harnesses—complete with woven-cloth insulation—for antique and collector cars. Check the suppliers list, pages 152-153.

In the case of a late-model car—up to 10 years old—the car dealer may be able to special-order the harness from the factory. Depending on the complexity of your car's optional electrical equipment, you may find a replacement harness at an auto wrecker. But be aware that you may be buying a new set of electrical problems. Also, the harness will be difficult to remove and will probably be filthy.

To make a replacement harness, follow the procedures explained in the chapter on constructing a harness from scratch, starting on page 100. In this case, you can use the original harness as a pattern for constructing the duplicate. Make sure the wire gages, insulation colors, terminals and connectors match the original.

To remove the original harnesses, disconnect the battery. Then systematically disconnect each connector and label it. You can use small paper tags with an attached string tie, available from stationery stores, or make your own labels with masking tape and a felt-tip pen, or whatever else works for you. Just make sure every wire is correctly labeled.

If you're replacing more than one harness, remove and replace each harness in the car, one at a time. Construct the first reproduction harness and install it in the car before removing the next original one. This keeps your work area un-

cluttered and lessens the chance of making wrong connections.

Using the original harness as a guide, construct a simple fixture, page 106. To make the fixture, use a sheet of plywood or whatever, with nails to form channels for each branch of wires. A sheet of pegboard with golf tees also works well.

Lay the harness on the fixture and form the channels in which the wires will run. Label all branch points on the fixture so you can identify them later. Then, remove the original harness from the fixture and lay it alongside the fixture, with each branch in its corresponding position.

Now, carefully unwrap the original harness, so you can see how each individual wire is routed. Use plastic ties or garbage-bag twist ties to keep the wires bundled, but with the wires visible.

Using the original harness as a guide, install the new wires in the fixture so they correspond with the wires in the original harness. Carefully match wire gages and colors, if possible. Compare conductor diameters—not insulation diameters—by eye. If you're not sure, err on the large side.

When measuring and cutting wires, allow 3 inches of excess wire at each end that will have a terminal. Before removing the harness from the fixture, label each branch exactly like you labeled the original harness. Leave the labels on the original so you can refer to them while installing the replacement harness and terminals. When all the wires are in place, install the appropriate harness wrap, tape or plastic ties, depending on the final effect you want.

Install the new harness in the car, referring to the labels to position all the branches. Secure the replacement harness to the car in the same way as the original. Be sure to eliminate all possibilities of binding, chafing or burning.

Attach terminals to the replacement harness, using the original as an example. Trim excess wire as necessary for a good fit. New terminals can be purchased from a well-stocked auto supply, car dealer or one of the suppliers listed on pages 152-153.

Although it doesn't look good, you can, if necessary, cut the terminals from

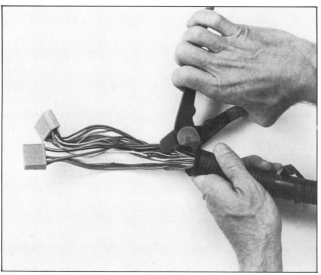

When making new harnesses, if you can't find original OEM connectors, you can cut them from old harness or one salvaged from auto wrecker. Leave several inches of wire to facilitate splicing to new harness. Photo by Jim Barrett.

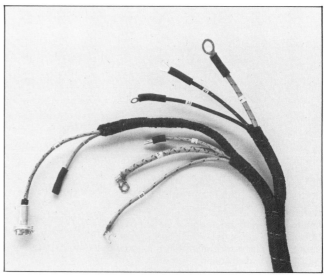

Authentic repro harnesses for antique and collector cars come with original braided-cloth wire insulation and loom, or with plastic insulated wires covered with cloth to simulate originals. This example from Harnesses Unlimited (see suppliers list) has exact matches on terminals and tracer colors, labeled wires and accompanying schematic. They also offer authentic harness materials for the do-it-yourselfer. Photo by Jim Barrett.

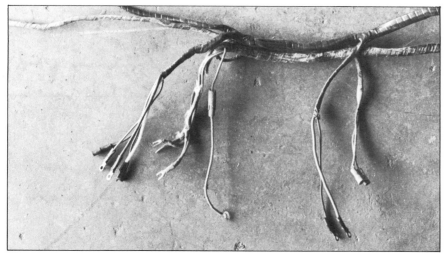

When you get your repro, lay it on top of original to make sure all branch points and wire lengths are correct, and all terminals match. A purist might send this new harness back because ring terminals were substituted for spades, but rings actually provide a better connection. Photo by Jim Barrett.

the original harness, or from one salvaged at an auto wrecker. Leave several inches of wire attached to each terminal. Then splice the cut ends to the corresponding wires in the new harness, using butt connectors or solder and shrink wrap.

Finally, connect the terminals to the components. Repeat this process for each harness you're installing. Connect the battery and test the circuits.

Installing a Reproduction Harness— Numerous suppliers can provide replacement harnesses for practically every car made—domestic or imported, collector or antique. This is the best way to go if you're restoring a show-quality car. Many of the terminals used on older cars

are no longer readily available. Neither is the lacquered, cloth-covered wire. Antique car clubs now accept modern *interpretations* of original wiring. These wires are covered with plastic insulation, and are then wrapped in a woven-cloth covering that simulates the originals. Suppliers are listed on pages 152-153. Also refer to specialty publications such as *Hemmings* and *Collector Car* magazines for names and addresses of others. You can often get additional references from members of various car clubs in your area.

Installing a ready-made reproduction harness is essentially a matter of fastening it into the car and plugging in the connectors. A word of advice, though. Before you install the reproduction, make sure it actually matches the wiring in your car. An easy way to do this is to lay the repro on top of the original on your work table to make sure all branches are correct and all terminals match.

Most reproduction manufacturers do accurate work, so if there's a mismatch, it's likely that your car's original harness was modified by an earlier owner. The repro manufacturers depend on word-of-mouth recommendations, so they want to be sure you're satisfied. So, don't hesitate to call them if you find an apparent discrepancy.

REPAIRING & REBUILDING COMPONENTS

The information in this section pertains primarily to collector cars, the parts for which are difficult to find. Cars newer than the '40s can usually be serviced by installing new or rebuilt components available at a car dealer or parts store.

No one book could possibly describe how to repair and rebuild every electrical component on cars that will eventually be restored. But specific repair manuals are available for most components. You may find the manual you need in *Hemmings,* the local library or at a local car club. However, there are some general principles that apply to all cars. These are discussed below.

Expendable Parts—When restoring or rebuilding a component, replace all expendable parts, such as light bulbs, fuses, brushes in motors and generators, and drive belts. Also, remove contaminants such as corrosion, oxidation, oil film, dirt, moisture, metal particles and expended lubricants. Painted components should be cleaned, primed and repainted.

Cleaning—In cleaning electrical components, use solvents that won't damage materials in the component. For instance, avoid acids, caustic solutions and water-based cleaners. Use solvents specifically made for electrical work, such as contact cleaner—generally available in aerosol cans—and lubricant/water dispersants such as CRC or WD-40. The strictly mechanical parts of electrical components can be carefully cleaned in degreasing compounds such as Gunk, as long as the part is rinsed and air dried before reassembly.

Protective Coatings—The windings in a motor, relay or coil are sometimes coated with clear lacquer to prevent moisture penetration. If you discover chips in the coating, they can be repaired by brushing or spraying. Just make sure the windings are clean so you don't seal in any corrosion.

Off-Car Testing—Components can be tested for continuity, shorts, opens and correct operation after they're removed from the car. Refer to the appropriate shop manual for test hookups. Make sure the components are correctly grounded while testing—an incorrectly grounded component may test as defective. Also, check your test instruments. Nothing is more embarrassing than condemning a component on the basis of a burned-out test lamp or ohmmeter with a dead battery!

FIX OR REPLACE?

This depends on the type of component, extent of damage and relative costs of repair versus replacement. It's usually easier and cheaper to replace minor components than to repair them, unless replacements are unavailable. Each component must be taken on an individual basis. Some specific components are discussed here.

Switches—If a switch is bad, replace it. The cost of repairing a switch is far greater than replacing it. Cleaning a glazed or corroded switch is also impractical. For newer cars, you may find a serviceable replacement at an auto wrecker. If you're restoring a rare or antique car, refer to the component suppliers, pages 152-153, for replacement sources.

Lamps & Sockets—Lamp sockets in modern cars—1970 to present—are fairly non-interchangeable because the socket is often integral with the lamp housing-and-lens assembly. But, the contacts—the parts most subject to wear and corrosion—are often soldered to the connecting wire, so they can usually be replaced.

Most lamp sockets for older cars were more universal in application—they fit a number of different car models of various manufacturers. Check some of the larger supply houses listed in the suppliers list. AuVeCo, Cole Hersee, or Terminal Supply, for instance.

Lamp bulbs are fairly interchangeable because their bases are standardized. The most common base for automotive use is the *bayonet.* For the more esoteric lamps, consult the specialty houses listed in *Hemmings* or check the suppliers list.

Instruments—It's best to replace defective electrical gauges. Check the sending unit first, to make sure it isn't the cause of the problem. A fairly brisk gauge business exists in the collector-car aftermarket. For newer cars, an auto wrecker may be your best bet for factory-original gauges.

Electric Motors—It's usually less expensive to buy a replacement motor. But if you can't find one, the original can generally be rebuilt. Bearings are fairly standardized, and rewinding can be done by a competent motor-service shop. Often, all a motor needs to restore it to life is to have the brushes replaced, the commutator polished or turned, and the insulation undercut. The latter two operations also are best left to a motor-service shop.

Specialized Repairs—Such components as generators, voltage regulators and starter motors are best repaired by specialists. With the exception of brush replacement, specialized equipment is required for testing and repairing these components.

Depending on the age of your project car, rebuilt components may be available. If they are, rebuilts will probably cost less than having the original rebuilt. Just be sure to get an exact replacement. Of course, replacements may not be available for collector cars or newer limited-production cars, so rebuilding the original component may be your only recourse.

Converting From 6 to 12 Volts

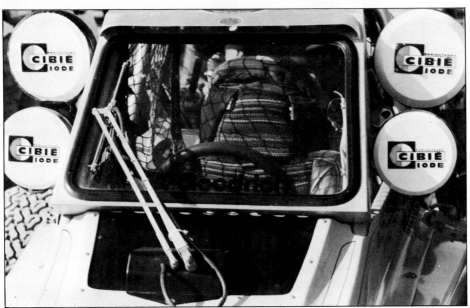

Original 6-volt system was barely adequate for stock pre-1967 VWs—no way would it handle this much candlepower on an off-road buggy! You'll probably want to replace 6-volt system anyway, because original system will probably be showing its age—and most aftermarket electrical accessories are 12 volt. Photo by Tom Monroe.

For model year 1967, Volkswagen changed from a 6-volt to a 12-volt system. It was a milestone year because the change eliminated cold- and hot-start problems caused by low voltage. Gone were the gremlins that caused headlamps to glow weakly and windshield wipers to slow to a snail's pace. It was an overdue case of catch-up, because American auto manufacturers made the switch in the mid-50's.

Since 1967, some owners of older VW's and American cars with 6-volt systems have sought the best way to convert their cars to 12 volts. Is it hard? Expensive? Worth it? Opinions vary. The purpose of this chapter is to present all the facts so you can decide for yourself.

Although this chapter describes how to convert a 6-volt system for a VW, the theory and general practices can be applied to any 6-volt system. The actual hardware and procedures may vary,

though. A knowledgeable parts-counter person can help you with the specifics.

WHY CONVERT?

Generally, there are two reasons for wanting to convert to 12 volts, and two reasons for not wanting to. The reasons for not converting are simple, so these will be discussed first.

If you're restoring a collector car back to original condition, you have no choice. You must remain with what is original on the car. Most American cars with 6-volt systems fall into the restoration category, unless you're building a street rod or similar car with an engine based on a 12-volt system.

The second reason for staying with a 6-volt system is the system is working satisfactorily and you don't think the benefits warrant a conversion. In other words, if it ain't broke, don't fix it.

On the other hand, you may want to

convert for one of two reasons, maybe both—*improved electrical-system capacity and reliability,* or *compatibility with 12-volt accessories.* A common example is the addition of driving lamps to a VW-based off-road car. For additional information on electrical systems for these cars, get HPBooks' *Baja Bugs & Buggies.*

Improved Capacity & Reliability— The 6-volt systems on older cars were usually well-engineered, and adequate for their intended task—barely. Most of these had little, if any, reserve capacity, so any added burden severely reduced the efficiency of the entire system.

On VW 6-volt systems, two things limit electrical performance. The wire gages used, though not actually stingy, are smaller than on any other 6-volt car. In addition, electrical paths in the circuits are quite long. With the battery under the back seat, current must flow forward to the instrument panel, then all the way back to the taillamps, ignition coil, starter solenoid, and so on.

As explained in chapter 2, resistance in a circuit is related to both diameter and length of a wire. In this respect, the VW's wiring is barely adequate for a 6-volt electrical source.

In addition to the above shortcomings, as a car ages, corrosion and oxidation attack terminals; switch contacts glaze over, losing some of their current-carrying efficiency; brushes and commutators in electric motors glaze and wear, and the bearings or bushings begin to gall, and fuses corrode in their holders. Taken alone, none of these may be enough to seriously affect a circuit. Added together, they conspire to drop available voltage below the usable minimum. As a result, lights dim and motors slow.

Remember that the last year VW used a 6-volt system was in 1966. Because you're dealing with a car at least this old, the electrical system will probably show its age.

It's a common oversight to ignore the unseen, forgotten second half of any circuit—the ground path back to the battery. Corrosion buildup here is just as devastating to electron flow as it is on the feed side. Maybe more so, because it's the last place most people think to look. Remember that the electrical components use the metal parts of the car—engine block, body and frame—as the return path, or ground.

To illustrate this point, I'll trace the ground path of the starter motor in a VW 6-volt system. Current leaving the starter motor flows into the transaxle through the starter-to-transaxle mating surface and the attaching bolts. Then there's supposed to be a transaxle-to-floorpan strap to carry current into the floorpan. In an older VW, the strap may be disconnected, corroded or missing. If it is, current will seek an alternate path, which is usually through clutch, accelerator and choke cables.

Under the high current flow of the starter motor, these cables have been known to fuse from the resulting heat.

From the transaxle strap, current flows along 2 feet of the floorpan to the battery-ground-cable attaching point. With so many junctions to flow through, it's no wonder that ground paths are the hidden, overlooked cause of many electrical problems in the VW system.

This illustrates how the VW 6-volt system, which is barely adequate to begin with, can eventually become entirely inadequate. Referring to the starter circuit again, it's not uncommon to find a 1.5-volt drop in an older car, compared to 0.2—0.3 volts in a new one. This 1.5-volt drop represents a full 25% drain from available voltage. Compare this to the same 1.5-volt drop in a 12-volt system, which is 12.5%, or exactly half that of a 6-volt system.

So, the advantage of the 12-volt system becomes clear. A smaller percentage of the available voltage goes toward overcoming corrosion and other unwanted resistances, leaving a greater percentage to operate the loads.

As an aside, think of the advantages of converting to 12 volts *and* cleaning up all the age-induced resistances that have built up over the years.

Compatibility with 12-Volt Accessories—Other than replacement parts for original equipment, there's little market for 6-volt electrical components. Certainly not enough to encourage manufacturers to offer their latest accessories in a 6-volt version: stereos, halogen lamps, power goodies, and so forth.

There are several compromise methods of using 12-volt accessories in a 6-volt car. And parts are available, although they're uncommon. They include a series-parallel installation of two 6-volt batteries, or the use of a *power inverter*. However, both methods are much less desirable than a straight 12-volt conversion, and are not recommended.

IMPROVING A 6-VOLT SYSTEM

The most telling argument against converting from 6 to 12 volts is that a well-maintained 6-volt system *will* work, provided you don't add more components than the system was designed to handle. All it takes is determination, some time and some work. And fixing what's already there will cost you less than changing over.

You know the basic symptoms of 6-volt illness, but to remedy them you need to know something about the causes.

From Ohm's law, you know that amperage is proportional to voltage; if you lower the available voltage, you lower current flow, or amperage. And, amperage is inversely proportional to resistance—add resistance and amperage drops. Further, you know that amperage is what makes electrical components work. From this, you can deduce that if you decrease resistance in the system by finding and eliminating voltage drops, more amperage will be available to operate the electrical components in the system.

Diagnosing Voltage Drops—Each electrical circuit in your car has several basic components—wire, fuse, switch, load—joined by terminals and connectors. Most auto circuits have all these components wired to each other in series, so think of each component as being one in a chain of components. You know there's supposed to be a *voltage drop* across the load because that's where the work is done.

But, if there's any corrosion or other undesired resistance in any terminal, the resistance will cause a voltage drop in that part of the circuit, leaving less voltage for the intended load. A voltmeter can be used to detect and isolate voltage drops in a circuit. To restore the efficiency of your 6-volt system, you'll need to find and correct these unwanted voltage drops.

There are two methods you can use: the *cumulative voltage-drop method* and *individual voltage-drop method*. Neither is necessarily better than the other. Before explaining them, though, I should emphasize that you can check for voltage drops only in circuits that are carrying current. This means the circuit must be turned on and the load functioning. If your battery is not fully charged before starting these tests, it may discharge before you're done, so first make sure the battery is fully charged and in good condition.

In the *cumulative voltage-drop* method, you're going to test the circuit from end to end, noting the increase in voltage drop as you go. To begin, firmly attach a long jumper wire to the battery negative post and attach the other to the voltmeter's negative probe. The jumper wire should be long enough so the meter and positive probe will reach any location on the car.

To test a circuit, first turn it on. Set the voltmeter to a scale that has 6 volts near the top, such as the 10-volt scale. Then begin by touching the battery positive *post* with the meter's positive probe. This gives you a *baseline voltage*. Write down the baseline voltage so you don't forget it. As you test each circuit, each reading you get as you move farther from the battery will be progressively smaller than baseline voltage. Be as accurate as possible, preferably to the tenth of a volt (0.1 volt). Any individual reading more than 0.2—0.3 volts less than baseline voltage indicates an excessive drop.

After establishing baseline voltage at the battery positive *post,* the first probe involves touching the battery positive *cable terminal.* If there's excessive corrosion between the post and terminal, the voltage reading will be significantly lower than the baseline. Clean the terminal and recheck it.

Now move on to the next junction in the circuit you're testing. On most cars, this would be the positive battery-cable terminal at the starter solenoid. Follow your chosen circuit all the way to the input terminal of the load, testing each junction in between. Check each terminal along the way.

As you go, you'll notice that your meter readings decrease as you move farther from the battery. This is to be expected because of resistance in the wire, as well as a small, unavoidable amount of resistance in each terminal-to-terminal interface. What you're looking for is one reading that's significantly lower than the trend. When you see one of these markedly lower readings, you've found a voltage drop that needs to be checked out. Remember that no meter reading should be more than 0.2—0.3 volts below the baseline.

If the trend-breaking reading you've found is caused by a corroded connection, cleaning it will probably cure the problem. If it's a switch, it's best to install a new one—glazed or corroded switches are difficult or impossible to clean permanently.

The *individual voltage-drop* method also involves testing a circuit that's turned on. But in this test, you're not looking for cumulative voltage drops. You'll test each terminal or component separately, looking for individual voltage drops. The total of all individual drops will add up to the cumulative drop.

The individual-drop method is not any more accurate than the cumulative method, nor is it any easier. If it has an advantage, it's that you don't need to compare readings with a baseline, and there's no subtraction required. Each reading is a direct voltage-drop indication.

Begin by turning on the circuit. Then find the connection you wish to test first. It doesn't have to be at the battery, although starting there is a good way to keep the test organized and systematic. Touch the positive voltmeter lead to the battery positive post and the negative meter lead to the cable terminal. If there's excessive resistance between post and terminal, you'll see a voltmeter reading. Because these readings will be small, use the smallest scale available—usually the zero-to-one scale. Incidentally, if you ever get the meter leads reversed, you won't damage anything—the meter will simply attempt to register below zero.

When you're testing a male/female spade-terminal connection, probe the terminal closest the battery with the positive meter probe. This will usually be the female one. Probe the other terminal with the meter negative lead. If there's a voltage drop between the male and female terminals, there will be a reading on the voltmeter.

If the reading is excessively high—0.1 volt is the maximum voltage drop for any one terminal—disconnect the terminals and clean them. Sometimes, just the act of sliding the terminals is enough to reestablish good continuity. That's why you must make the test before disconnecting anything. After cleaning, retest the connection.

Some components have a threaded stud for each wire to connect to. The wire has a ring terminal that's slid onto the stud and is retained by a nut and lock washer. To test such a connection, probe the ring terminal with the positive lead and the end of the threaded stud with the negative lead.

A large voltage drop indicates either corrosion or the lock washer is not getting a good bite on the ring terminal. Sometimes the problem can be cured by the simple disassembly/assembly routine, but go the extra distance and clean all parts carefully before reassembly and replace the lock washer.

You should expect a large voltage drop—close to that of the battery voltage—when you probe the input and output terminals of the load while it's operating. If the load is self-grounding—that is, it grounds through its mounting screws—probe the input terminal and the head of a mounting screw.

Finally, probe the head of the screw and the battery negative cable-to-floorpan attachment. This checks the voltage drop across the component's mounting screw and floorpan.

Check all junctions in the return path, including ground straps and the battery-negative cable attachment. On a bolted-together car such as a VW, the return path for some self-grounded components relies on the bolts and their metal-to-metal contact between separate body parts. A component mounted on a bolted-on fender is a good example. If a component attached to a bolted-on part, such as a taillamp on a fender, does not work correctly, check for corrosion where the parts bolt together.

Conical-end fuses used in most VWs are built-in corrosion collectors because the contact area is marginal. Twisting the fuse in its terminals usually reestablishes contact. Also, you can add a self-diagnostic feature to fuse panels that readily tells you when a fuse is not working correctly, page 36.

Internal contacts in switches can glaze over after prolonged use, leading to intermittent problems. The ignition switch is most susceptible because it is used the most. As mentioned earlier, defective switches should be replaced because cleaning the switch contacts is impractical.

As you make your tests, check the condition of the wiring and components. If you're dealing with the original 6-volt system, remember that it is at least 20 years old, and the wires will probably be showing their age. If you haven't done so already, read the chapter on rewiring to

Ford relay is used to bypass long circuit between battery and starter in 6-volt system, providing as much juice as possible directly to starter. See diagram at right. Photo by Jeff Hibbard.

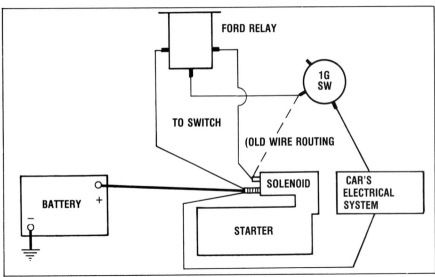

Wiring diagram for installing Ford-type relay. Mount solenoid close to starter.

determine if wires need to be replaced. For information on repairing and replacing worn components, see page 127.

HARD-TO-START CARS

Older VWs are known for their reluctance to start during cold weather, and often in hot weather, too. The cure for both problems is adding a *relay*. But it helps to understand the cause before making the correction.

A cold battery provides far less amperage than a warm one, as discussed on page 17. The VW starter solenoid requires fairly high amperage for operation. Current must flow from the battery forward to the starter switch and back to the solenoid, which is a great distance. The 6-volt battery, combined with added resistance due to the length of wire in the solenoid circuit, conspire to starve the solenoid, so it may not operate in extremely cold weather.

In hot weather, the same long wire is again the culprit. In this case, though, wire length plus increased resistance caused by heat combine to reduce current flow to the solenoid. So, on hot days, it may not operate either.

The solution is to add a *relay* to feed battery voltage through a shorter path that leads directly to the VW starter solenoid. A 6-volt starter relay for a 1945-55

Ford is used for this purpose. You'll use the original ignition switch-to-solenoid wire to activate the new relay. The accompanying diagram shows the correct hookup.

Mount the Ford relay as close to the VW starter solenoid as possible to keep wire lengths short. Be sure the relay is sufficiently grounded. It uses the mounting ears for ground conductors. Note that a 12-volt Ford relay will solve the same problem in VW 12-volt systems.

6-TO-12-VOLT CONVERSION

This section describes how to convert a VW 6-volt system to 12 volts. The basics, however, apply to any 6-to-12-volt conversion. Electrons follow the same rules regardless of brand names. It assumes that the car is otherwise stock. For instance, if you've installed an engine based on a 12-volt system into your VW, you can ignore the part about installing a 12-volt generator and coil, because the engine already has them.

Required Changes—First, I want to clear up some misconceptions. You *do not* need to change any wires, fuses or switches. Nor do you need to change the fuel gauge. It is mechanically operated on 6-volt VWs. The 6-volt horn will work for quite a while in a 12-volt system—change it only when it gives up.

You *must* change the battery, generator, voltage regulator, starter solenoid, ignition coil, radio, all light bulbs and turn-signal flasher. You have a choice whether or not you want to change the wiper motor and starter motor—see pages 132-133.

Generator—There are several considerations in replacing the generator. In addition to the obvious voltage difference, you must also consider the diameter of the generator body. It dictates the type of *generator pedestal* you'll need. The simplest, but potentially most expensive solution, is to install a *Bosch GR-26* 12-volt generator—it fits the existing 6-volt generator pedestal. But, because it is from the early Porsche 912, it is difficult to find and is priced accordingly.

A second choice is to install both the stock 12-volt generator and the pedestal from a later model VW. Like the 6-volt generator, it is rated at 30 amp. Check with local auto wreckers for these parts. As good insurance, replace the brushes and have output checked before you install the generator.

Your third choice is to use the 12-volt generator and pedestal from a 1969-71 VW van. It puts out 38 amp, compared to the 30 amp available from VW car generators.

The hot tip, though, is to install an

Ordinarily, you don't need to change wiring when converting a VW to 12 volts. But if you're wiring a modified bug or buggy from scratch, wiring kits like this one from Bugpack make the job easier. It contains enough color-coded wire, fusebox and connections to wire any buggy to street-legal specs. Photo by Ron Sessions.

years, so try to get the wiring diagram for the motor you've chosen.

Some people have the mistaken idea that you can use the original 6-volt wiper motor with 12 volts simply by reducing 12 volts to 6. However, there's no such thing as an all-purpose 12-to-6-volt resistor, so the conversion isn't all that simple.

Here's why: Because *amperage* determines motor speed, this is what must be controlled. To do this, you must know how many amperes you're dealing with. Connect an ammeter in series with the motor when it's running on 6 volts. The reading should be about 1.5 amp. There's the tricky part. You must get 1.5 amp to flow when 12 volts is applied.

Using Ohm's law, the 6-volt system flowed 1.5 amp, meaning the motor offered about 4 ohms resistance.

$$R = \frac{E}{I} = \frac{6 \text{ volts}}{1.5 \text{ amp}} = 4 \text{ ohms}$$

Now apply Ohm's law to the 12-volt system, keeping amperes constant at 1.5.

$$\frac{12 \text{ volts}}{1.5 \text{ amp}} = 8 \text{ ohms}$$

You had 4 ohms in the motor, so you must add 4 more ohms to get 8. Obtain a 4-ohm resistor. Be sure its wattage rating is adequate.

Watts = volts X amperes
W = 12 X 1.5 = 18 watts

alternator. VW started using them in 1973 Beetles, and they're rated at 45, 50 or 55 amp. The 1973 and 1974 models used a *separate mechanical voltage regulator,* while later models used an *integral regulator.* Numerous aftermarket suppliers carry complete alternator-conversion kits, including alternator, pedestal, regulator, and sheet metal for the fan-housing. If you're looking for increased output, these kits are more costly but represent a smart buy.

Voltage Regulator—Regardless of the generator or alternator you choose, use the 12-volt regulator designed for it. *Do not* use the original 6-volt regulator.

Ignition System—All you need to do is change to a VW 12-volt coil. A 6-volt coil used in a 12-volt system will destroy the ignition points.

Windshield-Wiper Motor—There are several choices here. The simplest and worst is to leave the 6-volt motor in place. It will run twice as fast and will eventually burn out. How long that takes depends on your local weather and the condition of the motor to begin with. Also, the excessive speed often causes the wiper's park switch to fail, meaning you'll have no way to shut off the wiper motor. All in all, not a good idea.

However, you can disassemble the 6-volt motor and *install the 6-volt gearbox* into a 12-volt motor. Then reconnect the gearbox to the linkage. This leaves all the 6-volt linkage intact. Be sure to use the wiper switch that goes with the motor you install. The wiring for these motors varies a little between various model

If you're looking for increased output to run extra lights or other accessories, install an alternator—it will put out 45 amp, even at idle. Photo by Jeff Hibbard.

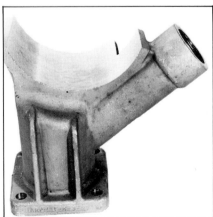

If switching from a 6-volt generator to a 12-volt type, you'll probably have to change the generator stand. This is a 12-volt stand. Photo courtesy of Bugpack.

So, you'll need a 4-ohm, 18-watt resistor—20 or 25 watts is better. If you can't find a single resistor, buy two and wire them in parallel. Add the wattage of each resistor to get overall wattage. To determine effective resistance (ohm value), use the parallel resistance formula, page 13.

Resistors generate a lot of heat when used this way. Keep them clear of any combustible or meltable components. It's a good idea to fasten them to a metal surface to provide a *heat sink* that will dissipate heat.

Although a viable solution, find an alternative to the resistor idea, if possible. It's too much of a cobble job.

Starter Motor—I've saved the hardest part until last. Here, you have two choices.

The cheapest and easiest way is to leave the 6-volt starter in place. The motor windings can be left as-is, but you must replace the 6-volt solenoid with a 12-volt one. If you don't, the solenoid is likely to overheat, especially if you crank the engine for any length of time. If the solenoid overheats and seizes so the starter drive is partially engaged, the flywheel teeth may be damaged. Also, leaving the 6-volt solenoid in place can cause an electrical fire in the solenoid/starter-switch circuit.

The consequence of putting a 6-volt starter on a 12-volt diet is that the starter will crank the engine much faster. If your car is hard to start, this may be good. However, it will also fling the drive gear into the flywheel with more than normal force. Over a period of time this will cause wear on the leading edge of the flywheel ring-gear teeth. Because VW doesn't use a separate ring gear on the flywheel, eventually the entire flywheel will have to be replaced. But, if you choose the only other alternative—using a 12-volt starter—you'll have to install a new flywheel anyway. So, unless you're building a race car, it really doesn't hurt to use the 6-volt starter until it fails, then convert to a 12-volt starter system.

The best, least-compromised starter conversion is to replace both the starter motor and flywheel with 12-volt components. Engine removal is necessary for installing them, which is why you may be tempted to leave the 6-volt starter and flywheel in place until one or the other finally fails.

The 6-volt starter is used with a 180mm flywheel; the 12-volt starter requires a 200mm flywheel. The installation of the 200mm flywheel is straightforward—consult the appropriate shop manual for details. A larger clutch must also be used with the flywheel.

Where it pilots into the transaxle case, the shaft of the 12-volt starter drive is smaller in diameter than the 6-volt one—0.43 in. versus 0.49 in. Bushings with smaller IDs are available. So are step-down sleeves. Replacing the bushing is the preferred method. Drive it out of the transaxle from the engine side, toward the front. Use a hardwood dowel and hammer to drive in the new bushing from the front toward the engine. Install the remaining components according to your shop manual. It all sounds much more difficult than it is.

Dropping resistors. Most of us have never seen one before. They can be used to drop voltage from 12 to 6 volts. Make sure you get one with the right specs. Photo by Ron Sessions.

Four bronze starter bushings. At bottom is stock 6-volt bushing. Right one is stock 12-volt bushing. At top is stock 6-volt bushing and special sleeve inside it to neck down bushing ID when switching from 6-volt starter to 12-volt one.

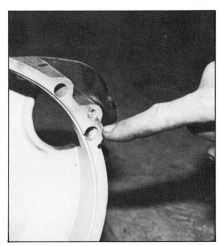
Starter bushing goes here. It is installed from opposite side. With engine installed, bushing can be removed only with the threaded bushing puller. With engine removed, bushing can be driven out. Photo by Jeff Hibbard.

Trailer Wiring

Ground for fender-mounted taillamp relies on metal fender attached to trailer frame, which in turn is attached to tow vehicle through hitch—a shaky ground at best. Better to run separate ground wire from lamp directly to good known ground on tow vehicle. Also note how wires are "hanging" unprotected beneath trailer. Bad wiring job is typical of that found on most trailers. Photo courtesy Trailer Boats Magazine.

Trailer wiring follows the same general plan as wiring a car, but it's much simpler. Usually, a trailer electrical system involves only lights—taillamps, brake lamps, turn-signal lamps and marker lamps. Rarely, a trailer will have back-up lamps. Some larger trailers also have electric brakes. This chapter shows how to install and wire the lighting system for a typical trailer. Special considerations for boat trailers are also covered.

This chapter will divide the trailer-wiring process into two parts: *trailer* and *tow vehicle*.

WIRING THE TRAILER

This is done before wiring the tow vehicle. Its electrical system will determine where and how you'll need to tap into the tow vehicle's electrical system.

Obtain the lighting equipment you want to install on the trailer. Because state laws govern trailer lighting, consult your state vehicle code to see what is required, before you buy the equipment.

Most trailer taillamps are designed for universal applications. They'll fit just about any trailer. They usually mount to the side of the trailer frame or an angle bracket attached to the side of the frame. In either case, to mount the lamp housing, drill one or two holes. Be sure to use lock washers on all fasteners for vibration resistance.

Trailer lighting is fairly weather-resistant to rain and road water. But, if you're wiring a boat trailer, you'll need to install special waterproof lamp housings that will operate underwater without damage. If the lamps don't have waterproof housings, water will leak in. This probably won't cause any permanent damage to the socket or wiring. But, if you happen to step on the brakes while the lamp is submerged—and you will, when launching a boat—the lamp will probably burn out from the hot bulb getting shocked by the cold water.

Trailer taillamps usually incorporate a single bulb that has two filaments—one for taillamp and one for brake/turn. This bulb is identical to those used on cars without separate turn signals. Some trailer lamps have a side-mounted lens that acts as a side reflector and side marker. Separate side markers are also available for the front (amber) and rear (red). Once you've selected and installed the lamp housings, you can wire the trailer.

There are two schools of thought regarding the ground for trailer lamps. Some contend that you can safely rely on the trailer frame as a ground, just like a car. Others say you should run a separate ground wire, similar to what you'd use on a fiberglass car body. I recommend using a separate ground wire that runs from the trailer lamps to a known good ground on the tow vehicle. Most trailer connectors include a ground terminal.

Here's a true story that illustrates what

If your trailer will include additional lamps or accessories, standard 6-pin connector gives you two additional circuits. Shown here is Six Way Trailer Connector, Part Nos. 11434 (male) and 11435 (female), Au-ve-co Products.

To protect harness, route it on inside of trailer tongue. Use clamps or frame clips as you would on car chassis.

can happen if you rely on the trailer frame to ground the lighting system:

I borrowed a friend's utility trailer to haul my kit car from the factory, a round trip of about 1500 miles. In winter. Clearly, a correctly operating lighting system was critical. I had wired my tow car according to the instructions given later in this chapter, and everything checked out OK. But when I hooked up the trailer, all kinds of weird things happened. The trailer taillamps worked, but dimly. The brake/turn lamps also worked dimly, except when the tail and brake lamps were on together, or when the tail and turn lamps were on together. When I stepped on the brakes, the taillamps went out and the brake lamps didn't light. When I operated the turn signals, the taillamp blinked on and off but the turn lamp did not light.

Here's a practical tip: *When electrical stuff works weirdly, look for bad grounds.* That's exactly what I found. The car's ground was connected to the trailer tongue through the harness. The tongue was connected to the trailer frame by a set of badly corroded pivots—the trailer has a tilt bed. The taillamps were grounded to the frame through their mounts. But, without continuity from frame to tongue, the lamps were not grounded.

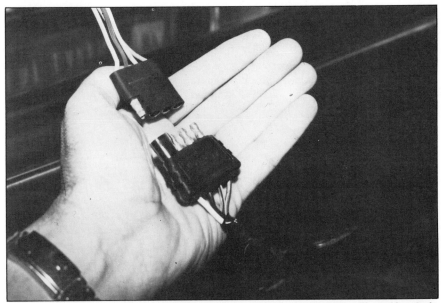

If your trailer has only taillamps and right/left turn and brake lamps, all you'll need is a 4-pin connector like this one, available from trailer-supply houses. They come with several feet of wire attached, in the four standard colors.

By attaching tail and brake lamp wires to female terminals, they are protected from short circuits. Exposed (male) ground pin can contact the car body without harm. This permits you to coil trailer harness under carpeting in trunk.

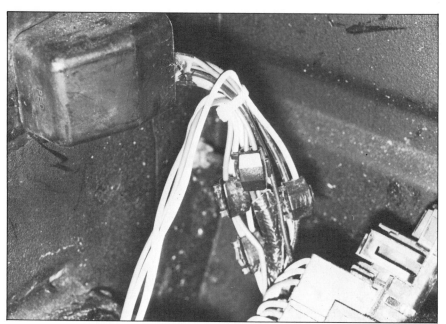

Tap into car's wiring at any convenient point. In this example, I was able to tap wires inside trunk, close to taillamps.

The taillamps were seeking ground through the car's brake lamps. And the trailer-brake lamps were seeking ground through the car's taillamps. When each system was operated separately, the trailer lamps lit, but dimly, because of the added resistance of the car's lamps. But when both systems were operated together, the trailer lamps had no ground, so nothing lit.

My solution was to run a ground wire from the trailer connector directly to the lamps, bypassing the tongue and frame, and the corrosion between them. It was a cold task, but necessary. So, don't rely on the trailer frame as a good ground, particularly on a tilt-bed trailer. Take the time to run a separate ground wire through the trailer connecter to a known good ground on the tow vehicle.

Connection to Tow Vehicle—For a conventional trailer system, you'll need only four wires between the tow vehicle and the trailer—ground, taillamp, right-turn/brake lamp and left-turn/brake lamp. These four wires come together at a connector at the front of the trailer tongue. A matching connector continues the circuits into the tow vehicle.

Trailer connectors are of two general types, each with four terminals. One type is flat, with all four terminals side by side. The other has the terminals in two rows of two, forming a square. Regardless of connector shape, the half of the connector that attaches to the tow-vehicle wiring has three female (hot) terminals shielded with insulation and one unshielded male (ground) terminal. The

For easy routing, trailer wire in the four standard colors—green, yellow, brown and white—is sold "stuck together" like this. Individual wires can be pulled or slit apart, if done carefully. Photo by Jim Barrett.

matching half that attaches to the trailer wiring has three unshielded male terminals and one shielded female terminal. When the two parts are plugged together, they can go only one way. And all male terminals are completely shielded by the connector bodies.

Wire colors for trailer wiring are fairly standardized. In fact, most trailer-hitch installers will sell you a connector with short wire leads already attached, in the four standard colors:

Green—right
Yellow—left
Brown—tail
White—ground

It's imperative that you install the correct half of the connector to the trailer and the correct half to the tow vehicle. Because you're going to be tapping into circuits that are normally live when the car is in operation, the connector should be installed so the live terminals are shielded against accidental short circuits. In other words, the female connector should be on the tow vehicle. Generally, the ground terminal on the female connector is left unshielded.

To install the trailer-half of the connector, first attach the appropriate connector leads to their respective circuits in the trailer. Use butt connectors and cover them with shrink wrap. You can test your trailer wiring with an ohmmeter or self-powered test lamp.

Touch one probe to the ground terminal; it's the *odd* one in the connector, and should be attached to the white (ground) wire. Then, probe each of the other three terminals in turn. If both the wiring and lamp filaments are OK, you'll see continuity at each connector. If not, trace down the open circuit and correct it. Then, temporarily remove the lamp bulbs and probe the three lamp-circuit terminals at the harness connector, two at a time. You should not see continuity between any two of them. If there is, you have a short circuit that must be corrected before hooking the trailer to the car.

WIRING THE TOW VEHICLE

The tow vehicle must be provided with a connector that mates with the one you installed on the trailer. I explained earlier that the connector attached to the car's

Be sure there's enough slack in trailer wire to permit trailer to turn in both directions without stretching wire. Keeping wire close to hitch ball helps.

wiring harness must have shielded terminals on all the live circuits. The ground terminal is generally unshielded.

The connector going into the tow vehicle will also have wires in the four standard colors. In fact, when you buy the connectors, they usually come in pairs, one connector installed at each end of a set of wires about 3 or 4 feet long. To install them, cut the wires in the middle, creating two separate connectors with several feet of wire attached to each.

If you can get a wiring schematic of the tow vehicle, use it as a starting point. But be careful. Factories have been known to assemble harnesses that have non-specified wire colors substituted for the ones specified on the schematic. This often happens when the factory runs out of a certain wire color. Instead of stopping production, they will substitute a color they have an abundance of. It may keep the line moving, but it can cause untold grief for the unwary technician later.

Find the rear lighting harness in the trunk of your car. On some cars, it may be underneath the trunk—bed of a truck—outside the car. On others, it may be behind the rear bumper. Tentatively identify these three wires: taillamp, right

turn/brake and left turn/brake. Now trace these wires toward the front of the car until you reach a connector. Use a voltmeter or 12-volt test lamp to verify each wire. Here's how.

Turn off all electrical systems in the car. Insert the test probe into the back of the connector in the cavity corresponding to what you think is the taillamp wire. *Do not give in to temptation to punch the test probe through the wire's insulation. This provides a place for corrosion to start later.* Ground the other test lead to a known good ground, such as a bare spot on the frame, or run a jumper back to the battery negative terminal. Then have a helper turn on the parking lamps only. If you see voltage at the chosen point when the lamps go on, you've found the wire for the taillamp circuit. Note its color and tracer, if any. If your car's harness is correctly assembled, it will match the shop-manual schematic.

Now, turn off the taillamps. Insert the test probe in the cavity corresponding to the left-turn signal. Have someone turn on the left-turn signal. You should detect voltage at the test point each time the lamps blink on. Identify the wire. Repeat for the right-turn signal. You have now identified the three wires to be tapped.

Trace the three wires to a location that's protected from the weather and from accidental contact. Under the trunk carpet is a good place. The place you choose will be your tap point. Hitch the trailer to the tow vehicle. Plug the trailer connector into the tow-vehicle connector. Route the leads to the tap point you've chosen. Be sure there is sufficient slack in the wires to allow the trailer to turn left and right without stretching the wires. Temporarily close the trunk lid to be sure the rubber weatherstrip will hold the harness in place without chafing. Open the lid and decide if the four wires should be shortened and cut.

Identify the car's taillamp wire and the corresponding brown wire from the trailer connector. Splice these wires together. Repeat for right and left turn. Finally, find a good ground for the white wire and attach it. You can splice it into a ground wire in the tow vehicle's harness or you can attach it to an existing screw in the car's body, using a ring terminal and serrated lock washer.

To attach a trailer harness connector to the car's wiring, use a snap splice connector. Insert feed wire into connector, then place wire to be added next to it.

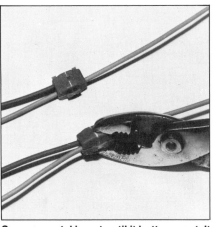

Squeeze metal insert until it bottoms out. It is designed to pierce the insulation of both wires, making electrical contact, without cutting the conductors.

It's perfectly OK to route trailer harness out of trunk between lid and weatherstrip. Just try to avoid trapping a nylon tie in weatherstrip; it could wear a hole in it.

HOW TO SPLICE A WIRE

I suggest using *snap-splice connectors* for this procedure. They're marked under various names, including tapping connector. The only tool needed is a pair of pliers, and you don't need to cut into the car's permanent wiring. Select a snap-splice connector of the appropriate wire gage. Open it up and be sure the metal slider is pulled out. Attach the connector to the appropriate tow-vehicle wire, using the straight-through channel of the connector. Insert the trailer connector wire in the dead-end channel. Snap the connector shut. Then squeeze the metal slider with pliers until it bottoms. Two slots in the slider straddle the two wires, slice through the insulation, and contact both wires. Now, snap the protective flap shut. For additional protection, you can tightly wrap the connection with electrical tape.

There's an alternate method you can use, but I don't like it and don't recommend it, because you must cut and strip the wires in the tow vehicle. This shortens the wires by about 1 inch and could put a strain on the splice.

Determine where you're going to put the splice. Cut the tow-vehicle wire and strip about 1/4 inch of insulation from each of the cut ends. Strip 1/4 inch of insulation from the trailer-connector wire. Slip an insulated closed-end connector onto the three wire ends together. Crimp the connector, using the correct-size die on the crimping tool. Tug on all three wires to be sure the crimp is tight. If not, pull off the connector and try again with a new one. For stress relief, wrap a nylon tie around the three wires close to the connector and pull tight.

Special Situation—Some newer cars, and many imports of various ages, use one set of taillamps for brakes and a separate set for turn signals. Often, the turn signal is amber, but not always. You cannot wire one of these cars exactly as described earlier, because the trailer uses only one lamp for both functions and you cannot power it from two sources in the car. So, you'll need an adaptor for this installation. Although all of these adaptors come with instructions, I'll briefly describe how they work.

Because the trailer still has the usual number of lamps, you'll use the conventional trailer wiring and the four-wire connector already mentioned.

Inside the car, find wires for the taillamp, brake lamp, left-turn lamp and right-turn lamp. The brake lamps will be on a separate wire from the turn signals. Now connect the leads from the adaptor box to each of the wires mentioned. Then, attach a standard four-wire connector to the other side of the adaptor. Route it to the rear of the car and connect to the trailer. The procedure is not very different, really.

Turn Flasher—Regardless of which type of turn-signal system your car has, you must replace the standard turn-signal flasher with a *heavy-duty* flasher. This is because the additional load of the trailer lamps on the circuit would cause a standard turn flasher to flash at an excessive rate.

The flasher unit is usually installed under the instrument panel. To find it, switch on the turn signals and follow the sound. Don't confuse the *turn* flasher with the *hazard* flasher: As discussed earlier, they look alike but aren't the same inside. Unplug the turn flasher and put it in the glove box for later use. Then, plug the heavy-duty flasher in its place. You can buy heavy-duty flashers at trailer suppliers. Or, buy a *hazard flasher* at your car dealer or auto-parts store. They are both the same unit. Both are designed to operate extra lamps.

While the trailer is in operation, periodically check lamp operation. The added load of the trailer lamps may exceed the car's fuse capacity. This could occur if you had the hazard flashers operating and simultaneously operated the brakes for much time, such as going down a long grade. The car's four turn-signal lamps plus the two on the trailer would remain on without flashing under these conditions, which could overload the hazard-flasher fuse. So, carry some spare fuses. *Resist the temptation to insert a larger fuse!*

Wiring a Race Car

All switches in instrument panel of Trans-Am Mustang are clearly labeled for instantaneous reference while driving. Race-car instrument-panel design is critical. Photo by Tom Monroe.

Wiring a race car is not much different than wiring a regular street car. But the variations, where they exist, can make the difference between winning and losing. Most race cars don't have complicated wiring systems, but these systems must work flawlessly, every time, under conditions that would literally shake a conventional wiring system to pieces.

The differences, then, are largely in the *quality* of the components and the car on which they're installed. It's not at all uncommon to find *aircraft grade* electrical components in a race car. The reason for using these costly components is the same for race cars as for airplanes—absolute reliability, uncompromised by price. Suppliers of race-car-quality components are listed on pages 152-153.

If you examine a typical race-car electrical system, you'll find wiring harnesses, terminals, circuit protection, switches, gauges and loads. These are discussed here, along with a few racer's tips tossed in where applicable.

By the way, most components described in this chapter can also be used on a street car, if you want to spend the extra time and money to install them.

WIRING

Race-car wiring is made of copper, covered by a protective insulation; no different from other vehicles. There are three specific points to be aware of when specifying wire for a race car.

Wire Size—You must use wire that's of sufficient gage for the *maximum* loads anticipated under race conditions. For example, in normal use, the ignition system may draw only a few amperes, maybe as high as 6 or 7 amp. But at the high engine speeds that race cars demand, the ignition pulses come very close together. This can cause a system to draw up to 30 amp at peak rpm.

As an example, if the electrical system in your race car uses 18-gage wire for the ignition, it could cause an excessive instantaneous voltage drop, resulting in poor ignition performance. This could be

prevented by using 14-gage wire. Of course, this is a hypothetical case. If you're not sure, temporarily install an ammeter in series with the ignition coil to determine maximum amperes when the engine is at high rpm.

A good race-car ignition, such as the MSD-6T, has separate leads for the power circuit and the on/off circuit. The power leads (B + and ground) will be 10—12 gage for minimum voltage drop at maximum current loads. The on-off function is controlled by a separate 18-gage wire drawing 1/2 amp, which is routed through the ignition switch.

On ignitions with only a single power lead and ground, all current is carried through the ignition switch. So, with these ignitions, you'll need a switch rated to handle the ignition's maximum expected current draw under racing conditions.

Wire Flexibility—Wire used in a race car should be as flexible as possible to help prevent failure from vibration. Wire flexibility is directly related to the *num-*

Here's a handy item for wiring vehicle. "Cart" can be moved to where it's needed. Photo by Tom Monroe.

ber of strands in the wire. For instance, 18-gage wire may contain seven strands of 26-gage wire, 19 strands of 30-gage wire or 41 strands of 34-gage wire. For best flexibility, select wire containing the maximum number of strands for its gage.

Wire Insulation—Be sure the wire insulation is capable of withstanding high temperatures. PVC wire insulation is rated from 80C to 115C (176F to 239F), depending on the plasticizers, stabilizers and sheathing used in its manufacture.

The insulation on wiring sold at conventional auto-parts stores has a maximum temperature rating of 85C (185F), unless otherwise noted. General-purpose wire used for race cars should have an insulation rating of 105C (221F). At this rating, the insulation will withstand the rigors of racing without breaking down, unless it accidentally touches a hot manifold or coolant line.

Insulation can be *UL, MILSPEC* or *CSA* rated. Wires with a 105C (221F) insulation rating include UL 1015 series wire, CSA TEW-105 series wire or MIL-W-16878D wire. MIL-W-5086A is an older MILSPEC wire, often denoted as *aircraft wire.* Also rated at 105C (221F), it has white PVC insulation with white glass braid and a clear nylon jacket. These wires are manufactured by major wire companies, such as Belden and Alpha. The Belden 8900 series and

Alpha 157D, 185D and 1870 series are examples of good, general-purpose race-car wire. Newark Electronics is a national distributor of Belden and Alpha wire.

Wire insulations with higher ratings include *irradiated polyolefin* at 125C (257F), Kynar at 125C (251F), Tefzel at 150C (302F) and Teflon at 200C (392F). Wires with these insulations are used for special situations, such as where wires must be routed near a turbocharger.

HARNESSES

Race-car harnesses must be durable and easily serviced. Durable means that they will perform their intended service under the anticipated operating extremes of the car. Easily serviced means that the pit crew can find, identify and replace defective components quickly and efficiently, under the time pressures of a pit stop. The way a harness is constructed can affect both of these requirements.

Harness Durability—Wires have weight. Before you laugh at this seemingly obvious statement, think about what it means in terms of a race car. First, race cars are built to strict weight requirements. Anything that does

not contribute to driver safety, vehicle reliability and winning, is either minimized or eliminated.

Obviously, wiring cannot be eliminated entirely, but it can be kept to a minimum. This means routing the wire bundles over the shortest practical path from one end of the car to the other.

Wire gage is another weight factor. In cars where the designer pares off a few pounds by using exotic metals and plastics for structural components, he would do well to consider the gages and insulation weight of the electrical wires as another potential weight savings. Naturally, each wire can be only so light without sacrificing reliability. So, select wires of the lightest weight possible without compromising reliability. MIL-W-5086A has a lightweight insulation and is a good wire to use when weight is of paramount importance.

Another weight factor of wire relates to *vibration.* Heavier components tend to vibrate more vigorously than light ones, and wire bundles are no exception. Again, you must consider the other aspects of reliability, but you should try to keep your wire harnesses as light as

Example of poor gauge positioning—angle makes them hard to read directly, also induces glare. Photo by Tom Monroe.

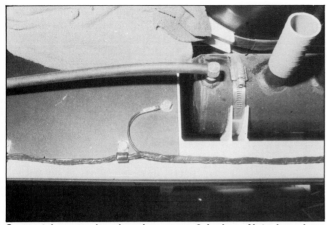

Support harness by clamping every 6 inches. Note here how branch point is close to a clamp for additional reinforcement.

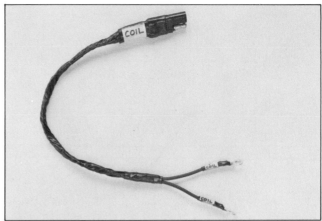

Coil lead made by race car electrician Frank Parker is example of good race-car wiring. He prefers harness wrap of nonadhesive loom tape to protect wires from dirt and grime. All leads are clearly marked with indelible ink on waterproof wrap-around labels for quick reference by pit crew during race. Aircraft-grade seamless ring terminals are used on coil leads. Photo by Jim Barrett.

Here, Essex BT-079 flame-retardant loom tape is being used as a harness wrap. Wrap wires tightly without stretching tape, overlapping by about one-half. Photo by Jim Barrett.

Because loom tape is nonadhesive, apply a piece of shrink wrap to keep ends from unraveling. Photo by Jim Barrett.

possible to reduce unwanted vibration. You can do this by using the smallest gage wire that will safely handle the maximum loads anticipated. Of course, the method of attaching the harness also has a lot to do with how much or how little the wires will vibrate.

Wire vibration is bad for two reasons: it leads to fatigue of the wires, and it can disconnect wires from components.

Always attach the harness securely over its entire length. Don't allow it to hang unsupported for more than 4—6 inches at any point.

Also, the weight of the harness should not place stress on any terminals. If it does, it could pull the connector loose at a crucial point in a race. Although you shouldn't have extra wire in the electrical system, allow just enough slack at each terminal to ensure that the harness is supported by clips or ties, and not by the component's terminals. If you expect any movement between the component and harness, put a small loop in the wire before it's attached. Such movement could occur, for instance, if the harness is attached to the firewall and the component is mounted on the engine. Wrap the wire around a pencil to form the loop.

Ease of Service—Service or repairs to the electrical system during a pit stop is affected by the way a harness is designed, constructed and installed.

Some racers like to bundle the harness with nylon ties rather than tape. The spaces between the ties allow you to see at a glance how the wires are routed. Replacement of single wires is also easier.

Other racers prefer to use a nonadhesive vinyl *loom tape* to bundle the harness. Although tape does make it slightly more difficult to trace wire paths and replace individual wires, it does have several advantages over nylon ties.

First, loom tape gives the harness a neater appearance than ties. Second, tape protects wires from dirt and oil. Third, a harness wrapped with loom tape will be lighter than one bundled with nylon ties.

Note use of ring terminals to ensure that these wires don't come unconnected at a critical part of a race. Use a dab of fingernail polish as added security in locking nuts in place.

And, because almost all wire failures occur at the ends of harness runs, it's rare that you'd need to remove the tape to make a repair in the middle of a harness during a pit stop. In any case, *do not* use standard electrical tape or other adhesive-type tapes for wrapping harnesses.

Each wire should be permanently labeled for easy identification. This makes it easier to troubleshoot electrical problems during a race, when time is important. You can label the wire with the name of the circuit, or with a significant code number understood by the pit crew. Durable wrap-around labels with clear-plastic overlays, available from Murphy Industries, are shown in the top right photo on page 141.

For marking labels, use a fine-point felt-tip pen with indelible ink—otherwise the ink will smear when the overlays are attached to the label. *Label every wire.* No matter how familiar you think you are with the wiring, it's easy to forget what each individual wire is for.

The top right photo on page 141 is a good example of how race-car wires can be wrapped and labeled. The two wires for a coil-harness lead are wrapped with Essex BT-079 flame-retardant vinyl

loom tape. FT-105 heavy-duty shrink wrap is used to keep tape ends from unraveling, also to reinforce terminal connections. Shrink wrap also makes a good, smooth surface on which to attach the labels described above.

The ring terminals used to make the harness vibration-resistant also make it difficult for the pit crew to disconnect a component quickly. However, this is a compromise that can be lived with. See the next section, Terminals.

There is no clear-cut preference for either clips or ties for attaching a harness to a race car. Clips have the advantage of looking more polished, but for a repair, the rivet must be drilled out. Also, it requires a rivet tool to install correctly. This is not critical during initial harness installation, but during a pit-stop repair, the extra work involved could cost the race. The advantage of the nylon tie, of course, is that it can be removed quickly with a diagonal wire cutter and can be installed without a tool.

On the other hand, there is some opinion that nylon ties can occasionally vibrate loose. If you use nylon ties, don't buy cheap ones. Look for ties with a small metal tab in the tie head. These are much less likely to vibrate loose than

those with plastic tabs. There are also ties that withstand high temperatures without melting, available from suppliers such as Murphy Industries.

TERMINALS

Many of the qualities to look for in wires for race cars can also be applied to terminals. They must be rugged, but not weigh too much. They must be able to resist vibrations and should not contribute to wire fatigue. Let's look at some of these qualities in more detail.

Terminal Design—Race-car electricians prefer terminals of a certain design. The terminal should be a ring rather than a spade, if at all possible. Obviously, when a wire attaches to a component, the design of the component is a factor in terminal configuration. That's why high-performance electrical components, and aircraft components used in race cars, usually have threaded stud terminals rather than the spade-type disconnect terminals.

Electrical systems on race cars may use soldered terminals, but the preferred type is the *solderless crimp connector.* At first thought, the soldered type would seem to offer superior electrical continuity and better retention of the terminal on the wire. Neither is true. A high-quality, correctly installed crimp connector is equal to a soldered one. In addition, it reduces a fatigue factor you probably never considered. When solder is applied to a terminal, some of the solder tends to *wick*—run—down between the strands of the copper wire. For the distance of the wicking, this makes the wire stiffer. But where the wicking stops, the wire is suddenly flexible. If fatigue from vibration is going to set in anywhere in the wire, this is the point where it will occur. A correctly installed crimp terminal covered with a piece of shrink wrap will eliminate the fatigue point.

The design of the connector barrel is also important. First, it should be *un-insulated.* This is because you're going to put shrink wrap on each terminal to reduce fatigue. Using insulated terminals would require a larger-diameter piece of shrink wrap, which would not shrink down correctly to support the connection.

There are three types of barrel

construction—seamless, silver brazed and seamed. The terminal barrel should be *seamless*. Terminals with silver-brazed barrels work well, provided they're correctly crimped, as described below.

Seamed terminals—the type found at almost all auto-parts stores—have barrels with open seams. Although suitable for conventional wiring jobs, they may not hold up in a race-car environment. Seamed terminals are usually of thinner construction than seamless types and have a tendency to lose their crimp.

Seamless and silver-brazed terminals are available through some aircraft-supply companies and national electronics distributors such as Newark Electronics. Regional distributors include Terminal Supply and R.S. Electronics. Refer to the suppliers list, page 153.

Crimping Terminals—Before attaching the terminal to the wire, slip a piece of shrink wrap onto the wire. See the section Installing Shrink Wrap for recommended type. Then strip the wire to the length required by the length of the terminal barrel. Slip the terminal onto the wire, making sure all the strands are completely inside the barrel. Then crimp the terminal using the appropriate crimping tool. Most terminals use the same kind of tool, although some have a variety of dies to accommodate different-size barrels.

As mentioned, seamless terminals are best, but if you use silver-brazed terminals, use a crimper designed for uninsulated terminals. When crimping, position the crimper's pointed die to the side of the barrel opposite the brazed seam. This precaution ensures that the die crimps the barrel without opening it up along the seam, which weakens the crimp considerably. A good crimping tool for this purpose is the Thomas & Betts model WT-2000, available at regional and national electronics distributors.

To be sure the crimp is adequate, tug on the wire. If it pulls out, install another terminal and crimp it tighter. Now you're ready to activate the shrink wrap.

Installing Shrink Wrap—First, buy a high-quality, thick-wall shrink wrap. Cheaper thin-wall shrink wrap will often split when heated enough to yield max-

The well-stocked race car electrician's terminal box looks like this. Note wide variety of terminals available. It helps keep everything neat and orderly, too.

imum shrinkage. A thick-wall shrink wrap also provides extra support for the terminal. Alpha FT-105 is one recommended kind.

You should have slipped a short piece of shrink wrap onto the wire before attaching the terminal to it. Now, slide the shrink-wrap tube back over the terminal, covering the barrel. If the terminal is a female spade connector, cover the body of the terminal, as well. In the case of male-spade or ring terminals, leave the body uncovered. Apply heat until the shrink wrap shrinks. Usually, a heat gun or a cigarette lighter is used for this purpose. Check manufacturer's instructions for the correct method.

Attaching Terminals to Components—Most components will have stud or screw terminals intended to accommodate a ring terminal on the wire. Generally, the stud will have a nut on it, or the screw will be furnished when the component is purchased. You may have to furnish the required washers. As added insurance against loosening, replace plain nuts with nylon-insert locknuts, called *Nylock nuts*. These nuts are thicker than plain nuts, so they require more stud length. Make sure the stud is long enough to pass through the plastic locking ring of

the nut. If more length is needed, you can remove the flat washer. *Do not* remove the lock washer—it prevents the terminal from twisting and placing stress on the wire when the nut is tightened.

The correct installation sequence for stud terminals is: 1. internal-tooth lock washer; 2. ring terminal; 3. plain (flat) washer; 4. nut.

The correct sequence for installing screw terminals is essentially the same, except that the items should be installed onto the screw *in reverse order* and *before* it is attached to the component: 1. plain (flat) washer on screw; 2. ring terminal on screw; 3. internal-tooth lock washer on screw; 4. screw attached to component.

The intent in each case is to place the internal-tooth lock washer between the ring terminal and component to provide a good electrical connection and prevent the terminal from twisting. Placing the flat washer beneath the nut or screw head provides a good bearing surface to prevent damage to the ring terminal.

Here's a *racer's tip* for screw and stud terminals. Use a dab of bright red fingernail polish on each connection to act as a mild thread lock. The polish locks the screws or nuts in place to keep them from

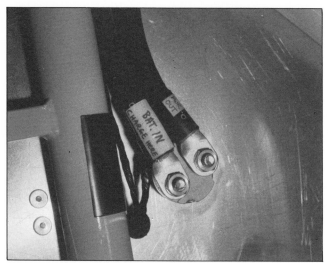

Battery cables were fabricated for this installation. Note permanent labels installed on them. These heavy wires lead to master cutoff switch on opposite side of panel.

This battery cable crimper is used to install cable terminals much like smaller ones used on wires. Except this one has to be squeezed in a vise or struck with a hammer to crimp terminal. The terminal is shown in place, but the cable was omitted for clarity.

vibrating loose. More important, you can tell at a glance if the fastener has turned, because the polish on the fastener won't line up with the polish on the surrounding area.

Battery-Cable Terminals—Terminals used on battery cables in race cars also have special requirements. First, it's often necessary to mount the battery a great distance from the engine for improved weight distribution, so custom cables are usually required.

Select cable terminals that can be *crimped* onto the cable. Don't use the type that clamp on—they could come unclamped during a race. The crimp terminal should be installed to the battery cable with the appropriate crimping tool. Obviously, the small plier-type tool you use for harness-wiring terminals won't do the job. There are two specific tools for this purpose: One is an expensive, heavy-duty plier-type tool; the other is a tool that is squeezed between the jaws of a bench vise or hit with a hammer to crimp the terminal to the cable.

Battery-cable terminals come in several configurations, depending on application. The *battery-terminal connector* attaches the cable to the battery; both side- and top-terminal kinds are available. The *lug connector* attaches the cable to the starter solenoid. The *T-type connector* is used to join two cables together at a battery post. They're needed when you have two batteries teamed together in the vehicle.

For battery terminals, there is a heavy-wall shrink wrap with a hot-melt sealant. It is available from Alpha Wire or NAPA automotive jobbers.

CIRCUIT PROTECTION

When the subject of wiring protection for race cars comes up, there's only one component that is even considered. That's the *circuit breaker*. These come in several configurations, but one kind is most suited for race cars. To repeat what was said earlier, circuit breakers are designed to react to excess amperage by

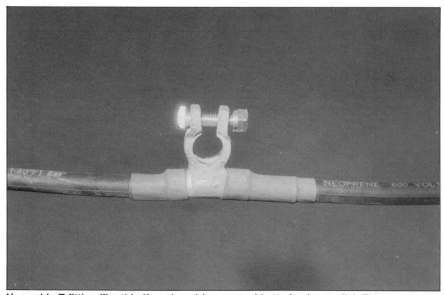

Use cable T-fitting like this if you're wiring several batteries in parallel. This terminal is suitable for top-post battery.

BATTERY DETAILS

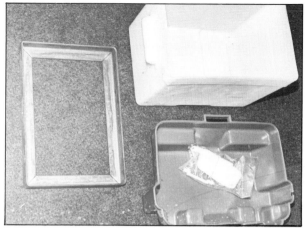

Battery box protects battery—and driver—during a crash. They're frequently used for off-road racing. Photo by Jeff Hibbard.

Battery box must be securely fastened to be effective. Photo by Jeff Hibbard.

Special frame was constructed to locate battery at rear of Trans-Am Mustang. This was done to improve weight distribution. Photo by Tom Monroe.

Some short-course racers are weight-conscious. They use light snowmobile or garden-tractor batteries for power. Photo by Ron Sessions.

overheating and snapping open. They don't melt like fuses do. The common circuit breaker remakes the circuit as soon as it cools. If the short is still there, it heats up, snaps open, and the cycle repeats. These are also called *auto-reset* circuit breakers. The auto-reset type is not the best for race-car use.

The preferred circuit breaker is one with a built-in toggle switch. Actually, it looks more like a conventional toggle than like a circuit breaker. But its operation is special. Normally, the driver turns the circuit on and off as desired, using the toggle switch. As long as there's nothing wrong in the circuit, the circuit breaker behaves just like a switch. But if an overload or short occurs, the built-in mechanism snaps the toggle switch open, breaking the circuit.

This type of circuit breaker is essentially a self-operating off switch. Unlike the auto-reset type, it will not turn itself back on. When the driver attempts to turn on the circuit manually, the circuit breaker will again break the circuit if the overload is still there.

High-quality toggle-type breakers are manufactured by Potter and Brumfield and Klixon. The former are available from national electronics distributors; the latter from aircraft-supply houses. Not surprisingly, these quality circuit breakers have two threaded studs on the back to receive ring terminals. Make sure the circuit breaker you select has a steel toggle. Breakers with plastic toggles have failed under racing conditions.

It's not only evident that using a circuit breaker as a switch means that the circuit breaker protects everything connected to it, but also that everything connected to it operates simultaneously. If you want to control something separately, you must add another circuit breaker. Or, you can

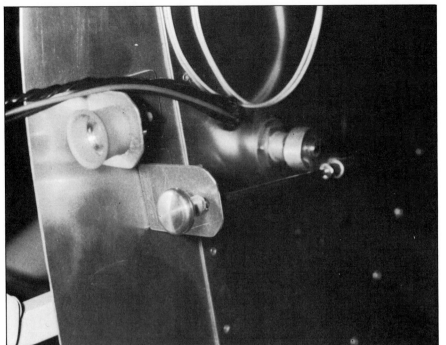

Location and type of master cutoff switch are determined by race sanctioning group. This switch is controlled by a linkage and knob.

Backup coil is common example of *redundancy* built into race cars. DPST switches wired to control a single circuit is another. Photo by Tom Monroe.

add an additional circuit breaker to act as a sort of "accessory" protection device. Because each component sharing this one circuit breaker has its own control switch, each can operate independently, provided the circuit-breaker toggle is switched on. Some of the components you'd put on this circuit include wipers, oil-pressure light and starter solenoid.

Sizing Circuit Breakers—In conventional street vehicles, breakers and fuses are sized slightly bigger in capacity than the current draw of the load. The idea is to protect the component from electrical damage. In race cars, circuit breakers are used to protect the entire electrical system—primarily the wiring harness—rather than individual components. So, the breakers in a race car should have much more capacity. During a race, the driver wants the component to operate as long as possible.

For example, if a fuel pump continuously draws several times its normal current, it will eventually die. But the driver is betting that it will last long enough to finish the race, which is more important than saving the fuel pump. If the circuit breaker keeps shutting off the pump, the race is lost.

Depending on the component, you can size breakers three to five times larger than the actual current draw of the component, but *do not* exceed the maximum amperage rating of the wires in the circuit. The circuit breaker *must* protect the harness.

SWITCHES

As you've already seen, it's common to use the circuit breaker's built-in toggle switch to control most circuits on a race car. But for those circuits having a separate switch, here's a *racer's tip* used by many wise race-car builders. From the discussion on switch types, page 30, you know that all you need to turn a circuit on and off is a single-pole, single-throw (SPST) switch. And, if you had a double-pole, single throw (DPST) switch, you could control two separate circuits with one switch activator. But you could also connect *both* poles of the DPST switch to the *same* circuit, providing two separate, independent paths for current to follow through the switch. If one pole should malfunction, the other pole would carry

Sealed switch. Dirt and dust are sealed out, keeping the switch contacts clean. Photo courtesy of Bugpack.

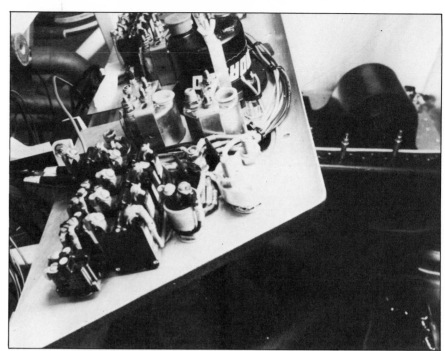

Race-car instrument panel is all business, and is laid out for easy service. It can be tipped out for access to wiring.

the current. This use of a switch is called *redundancy*. In a street vehicle, such redundancy isn't necessary. In a race car, it's extra smart. It can be especially aggravating to lose a race because of a failed switch.

Regardless of how many poles your car's switches have, or how redundant they are, there's one thing you should not do—use cheap switches. *Don't even think of using factory or aftermarket switches intended for ordinary street vehicles.* Use an aircraft-grade switch or an environmentally sealed MILSPEC switch. These switches cost much more, and most aren't very fancy, but they get the job done. And that's all that matters. See suppliers list, pages 152-153. For example, the TL-series switch, made by Microswitch, is an excellent aircraft-grade toggle switch.

One special electrical switch that's required by certain race-sanctioning groups is the *master cut-off switch*. The location is also determined by the rules of the sanctioning group, but essentially it's outside the body where it can be reached easily by a rescue person. The location of the master switch is usually identified by a painted symbol for further aid in finding it quickly. Being able to shut off the entire electrical system with one switch is critical in preventing fires in race-car accidents.

GAUGES

There are two schools of thought regarding race-car gauges. One holds that the gauges should be mechanical, and therefore outside the scope of this book. The other holds that all gauges should be electrical or electronic. Probably, both sides can make a valid argument to support their pet theory. Let's just say that if you choose electrical gauges, pay attention to the following precautions.

Auto Meter, a respected manufacturer of competition instruments, offers an extensive line of mechanical gauges for pressure, temperature and vacuum. But their tachometers are all electronic. And these instruments are to be found in a wide variety of successful competition cars. They also offer tachometers combined with a rev-control unit. And finally, their "mech/elec" tachometer combines a mechanical pickup driven by the distributor, but is electrically isolated from the ignition system. The pickup unit generates a signal to operate the electronic tachometer head.

Sending units for electrical temperature and pressure gauges work on the principle of providing varying resistance for the sensor terminal of the gauge. This means that gauge reliability is only as good as the sending-unit ground or pipe adapter.

Sending units are screwed into the engine block or head using the built-in tapered threads of the sending unit. If the threads don't make good contact with the engine, or if there's some pipe sealant in the threads, the gauge may not work at all, or at best, inaccurately. The same problem could occur if the sender or adapter is screwed into an anodized fitting, because the anodizing is usually aluminum oxide, which acts as an insulator.

Here's a hot tip. *Carefully* drill a small hole into the hex head of the sending unit's or adapter's threaded fitting. You must determine where the hole can be drilled, and how deep, to avoid ruining the sensing element inside. Then, tap the hole to accept a small machine screw. Finally, run a wire from a known good

Good idea: Dzus 1/4-turn fasteners allow quick access to backside of instrument panel. **Bad idea:** Protruding gauge bezels may look sharp and shade gauges from direct sunlight, but they would also *feel* sharp should driver to be thrown against console during a orash. Photo by Tom Monroe.

This oil temperature sender, mounted in the oil reservoir, has been drilled and tapped to permit installation of a separate ground wire. With wire in place, sender does not rely on contact through pipe thread on sender body.

ground to the screw installed in the new hole in the sender. Now, the sender will be adequately grounded regardless of contact made by the threads. Again, it's extra work, but such attention to detail wins races.

Warning Lamps—It's a good idea to provide warning lamps as back-ups for at least the oil-pressure and charging-system gauges. Off-road race cars often use warning lamps instead of gauges because the jostling doesn't permit interpreting gauge readings anyway! They'll also "grab" the driver's attention better than the needle of a gauge. And, the dust and vibration encountered in off-road racing can destroy the best of gauges.

For best visibility, warning lamps should be large. Round clearance lamps for trailers and RVs make excellent warning lamps. The ones shown in the photo are installed on the instrument panel of a competition Baja bug. Note how lamps are shaded for better visibility in daylight.

Adjustable pressure switches for oil-pressure lamps are available from suppliers like Stewart Warner or Moroso.

LOADS

The loads used in any specific race car are determined by the nature of the competition, the rules of the sanctioning body, the preferences of the driver and the car builder, and the requirements of the car itself.

To clarify this, let's suppose you're going to wire a car intended for Formula Vee racing. There are only a handful of loads on the entire car. The starter motor is conventional, the ignition is conventional, and the gauges are up to the individual team. No lights are needed or permitted. Electric fuel pumps are not permitted. From what this book has taught you about wiring already, you could probably wire such a car in your sleep.

LIGHTS

Cars intended for use in professional road rallying and off-road racing require an extensive lighting package. A typical setup would be a set of two high-output driving lamps and a set of two spotlights or fog lamps. Such cars require a beefed-

A good time to wire instrument panel is when you still have good access to it. Photo by Tom Monroe.

Trailer clearance lamps, shaded from sun, make hard-to-miss warning lamps on this competition Baja Bug. Stick at right of instrument panel is horn-switch acutator that can be operated with a slap of hand. Large knob below floods engine and driver's compartment with Halon-13 fire-extinguishing agent. Photo by Jeff Hibbard.

up charging system to handle the extra loads. This includes a high-output alternator and battery with a high reserve capacity. However, when wiring lights for a race car, perhaps the most important decision you must make is how you want to control the lamps.

Some racers prefer a special control board with toggle-type circuit breakers to control the lamps. Setups vary: one breaker for each lamp; one breaker for driving lamps and one for spotlamps, or one breaker to control a set of lamps on each side of the car. Which one you choose depends on the amount of circuit protection you want versus the number of switches you want to deal with during a race. The third configuration makes sense in racing—if a front corner of the car gets damaged, damaging the lamp circuit on that side, the lamps on the other side will continue to work.

The advantage of the toggle-breaker setup is that it doesn't require relays, which are just additional components that could fail during a race. Because toggle-type breakers combine the switching and circuit-protection functions into one unit, if they pop off due to a momentary overload, they can be immediately reset, unlike fuse-protected relays.

Relays—There are situations where a setup using relays may be preferable to the toggle-breaker setup just discussed.

Class 7 off-road racing truck relies on minimum instrumentation, resulting in simple, uncluttered dash. Light switches are mounted on panel between seats. Photo by Tom Monroe.

One is if you want to use low-amperage switches instead of breakers to control the lamps. For example, let's say a circuit with two 100-watt driving lamps draws a total of 15 amp, but the switch you've chosen won't handle that much current. Note that few standard switches are designed to handle over 15 amp, although several lighting manufacturers offer switches with their lighting kits that can handle up to 40 amp. If you choose to use a switch with a low amp rating, you'll need to control the lamps with a switch-actuated relay. The switch, fed by the instrument-panel circuit, carries only a small amount of current to operate the relay, which in turn, handles the load. Depending on the number of poles and throws in the switch, it can operate one or more relays in one or more circuits. One example is to use a relay for each lamp in a two-lamp circuit, with one switch controlling both relays.

The second situation is when excessive wire length between the lamps and the switches on the instrument panel would cause an unacceptable voltage drop in the system. Using a relay shortens the length of wire required to carry current to the lamps, reducing voltage drop in the wire. Using a heavier gage wire or one with a larger number of strands will further reduce voltage drop.

For example, assume the circuit for the 100-watt driving lamps uses 18 feet of 14-gage wire. Using the chart and instructions on determining wire sizes, page 23, you can calculate total wire resistance, which is 0.05 ohm. Using Ohm's law to calculate voltage drop, you get: R (0.05 ohm) X I (15 amp) = E (0.75 volt).

As a rule of thumb, voltage drop in the wire should not exceed 0.5 volt. Now, assume that by using a relay you can wire the same lamps with 8 feet of 12-gage wire. Applying Ohm's law, you get: R (0.014 ohm X I (15 amp) = E (0.216 volt), which is acceptable.

Relays may be necessary in some racing classes that require that all lights be able to be switched off by a single control. If you're using breakers, this could be handled with a master light switch, but could also be easily done by using a single switch to activate all of the relay coils in the headlamp circuits. In the latter

High-output driving lamps for off-road buggy are mounted as high as possible for optimum range. Lamp covers are a must for protecting lamps. Photo by Tom Monroe.

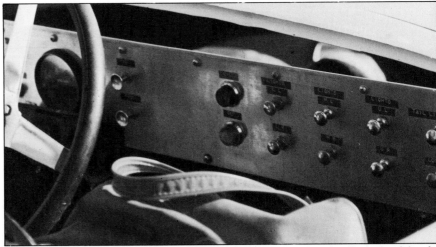

For off-road racing, driving lamps are operated individually or in pairs. Each circuit has its own circuit protection. Photo by Tom Monroe.

case, the switch would handle only the relay coil current (about 0.050 amp).

Other complex switching arrangements can also be handled easier with relays. If you use relays, each circuit still requires protection—an in-line fuse, fusible link or circuit breaker. If you go with fuses, your best bet is to use relays with built-in fuse holders, provided by the lighting manufacturer. Otherwise, you'll have to install in-line fuses to protect the circuits.

METRIC CUSTOMARY-UNIT EQUIVALENTS

Multiply:		by:		to get:	Multiply:		by:		to get:
LINEAR									
inches	X	25.4	=	millimeters(mm)		X	0.03937	=	inches
miles	X	1.6093	=	kilometers (km)		X	0.6214	=	miles
inches	X	2.54	=	centimeters (cm)		X	0.3937	=	inches
AREA									
inches^2	X	645.16	=	millimeters^2(mm^2)		X	0.00155	=	inches^2
inches^2	X	6.452	=	centimeters^2(cm^2)		X	0.155	=	inches^2
VOLUME									
quarts	X	0.94635	=	liters (l)		X	1.0567	=	quarts
fluid oz	X	29.57	=	milliliters (ml)		X	0.03381	=	fluid oz
MASS									
pounds (av)	X	0.4536	=	kilograms (kg)		X	2.2046	=	pounds (av)
tons (2000 lb)	X	907.18	=	kilograms (kg)		X	0.001102	=	tons (2000 lb)
tons (2000 lb)	X	0.90718	=	metric tons (t)		X	1.1023	=	tons (2000 lb)
FORCE									
pounds—f(av)	X	4.448	=	newtons (N)		X	0.2248	=	pounds—f(av)
kilograms—f	X	9.807	=	newtons (N)		X	0.10197	=	kilograms—f

TEMPERATURE

Degrees Celsius (C) = 0.556 (F - 32)   Degree Fahrenheit (F) = (1.8C) + 32

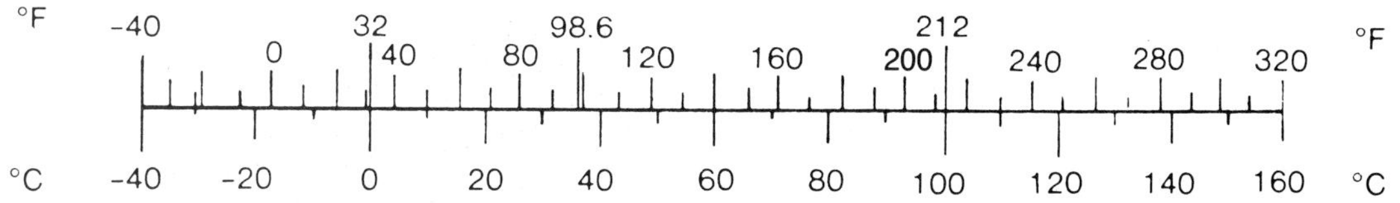

°F	-40			32		98.6				212					°F
		0	40	80	120	160		200	240	280	320			
°C	-40	-20	0	20	40	60	80	100	120	140	160			°C

ENERGY OR WORK									
foot-pounds	X	1.3558	=	joules (J)		X	0.7376	=	foot-pounds

FUEL ECONOMY & FUEL CONSUMPTION

miles/gal	X	0.42514	=	kilometers/liter(km/l)		X	2.3522	=	miles/gal

Note:
235.2/(mi/gal)=liters/100km
235.2/(liters/100km)=mi/gal

PRESSURE OR STRESS

inches Hg (60F)	X	3.377	=	kilopascals (kPa)		X	0.2961	=	inches Hg
pounds/sq in.	X	6.895	=	kilopascals (kPa)		X	0.145	=	pounds/sq in
pounds/sq ft	X	47.88	=	pascals (Pa)		X	0.02088	=	pounds/sq ft

POWER									
horsepower	X	0.746	=	kilowatts (kW)		X	1.34	=	horsepower

TORQUE									
pound-inches	X	0.11298	=	newton-meters (N-m)		X	8.851	=	pound-inches
pound-feet	X	1.3558	=	newton-meters (N-m)		X	0.7376	=	pound-feet
pound-inches	X	0.0115	=	kilogram-meters (Kg-M)		X	87	=	pound-inches
pound-feet	X	0.138	=	kilogram-meters (Kg-M)		X	7.25	=	pound-feet

VELOCITY									
miles/hour	X	1.6093	=	kilometers/hour(km/h)		X	0.6214	=	miles/hour

Notes

Notes

Glossary

AC Generator—An electromechanical device that generates alternating current (AC). Commonly called an *alternator,* it is usually belt-driven off the engine. It provides maximum output at relatively low rpm and is used on all modern cars. It requires a rectifier to convert alternating current (AC) to direct current (DC) required by auto electrical system.

Aftermarket Parts—Components that can be added to a vehicle after manufacture. These parts are often accessories, and are not to be confused with OEM (original equipment manufacturer) service or replacement parts.

Ammeter—An instrument-panel gauge used to measure flow rate of current into or out of the battery. It is calibrated in amperes for both charge (+) and discharge (-) rates, in ranges of 20, 30 or more amperes.

Ampere (amp)—A measure of current flow. 1 coulomb per second.

Analog Gauge—See *Gauge.*

Armature—In an electric motor, the coil of wires that are wrapped in a specific pattern. The coil surrounds the armature shaft or axle.

Battery—A cell or collection of cells that convert chemical energy into electrical energy. A reversible battery can be recharged by converting electrical energy into chemical energy. A battery has two poles—positive and negative. The amount of positive and negative charge is called *potential.*

Battery Ratings—Performance standards conducted under laboratory conditions to describe a battery's reserve capacity and cold-crank abilities. The amp-hour rating is no longer in widespread use. See *Cold-Crank Rating.*

Brushes—In an electric motor, the spring-loaded blocks that rub on an armature's commutator to conduct current through the armature. Some motors have multiple brushes, used to select several operating speeds.

Bulkhead Connector—An OEM device used to connect wiring inside the car body with wiring outside the body. Usually located at bulkhead or firewall.

Butt Connector—Solderless connector used to permanently join two wire ends together.

Circuit—An electrical path from source (battery or generator), through load(s), and back to source. A typical circuit consists of a battery, wire, load (lamp or motor) and switch. See also *Simple Circuit* and *Single-Wire Circuit.*

Circuit Breaker—A circuit-protection device, containing movable points that open if preset ampere load is exceeded. Some circuit breakers are self-resetting; others require manual resetting.

Clamp, Harness—Plastic or rubber-coated metal fasteners used to attach harnesses to car underbody, frame or panels. Attached by screws or rivets.

Clip, Harness—Metal fastener that can be pushed onto edge of chassis rail to retain harness. Also called a *frame clip.*

Closed-End Connector—Solderless connector shaped like a hat. Used to join two or usually three or more wires permanently together. Similar to wire connectors used in house wiring, but installed by crimping, rather than twisting.

Coil—1. Any electrical device or component consisting of wire loops wrapped around a central core. Coils depend on one of two electrical properties for operation, depending on application, either electromagnetism or induction. 2. Ignition coil. See pages 53-54 for description.

Cold-Crank Rating—The minimum number of amperes a fully charged 12-volt battery can deliver for 30 seconds at 0F without falling below 7.2 battery volts.

Commutator—In an electrical motor, a series of parallel metal bars attached to wire loops in the armature. Provides contact with brushes to bring current into armature.

Conductor—Any material that freely permits electron movement. Examples are copper, aluminum and steel. Conductors are all metals. The metal part of an insulated wire is often called the conductor.

Constant Voltage Regulator (CVR)—A device used to maintain a constant voltage level in a circuit, despite fluctuations in system voltage. CVRs are wired into some gauge circuits so voltage fluctuations won't affect accuracy of gauge readings.

Coulomb—An electrical charge of 6.25 X 10^{18} electrons, or the charge transported through an electrical conductor by a current

of 1 ampere flowing for 1 second. See *ampere.*

Current Flow—Movement of electrons in a conductor. According to tradition, electron flow is from positive to negative. According to the electron theory, flow is from negative to positive. This book follows the traditional theory. Current flow is measured in amperes.

Cylinder-Head Temperature Gauge—Indicates cylinder-head temperature in degrees F or C. Used to measure temperature of air-cooled engines. Temperature ranges encountered by gauge sensor are much higher than those in a water-cooled engine. Ranges between 100F and 600F (40C and 320C) are not uncommon, with 250—400F (120—200C) regarded as the normal operating range.

DC Generator—An electromechanical device that generates direct current. It is usually belt-driven off the engine. Because the DC generator requires high rpm for maximum output, it is no longer used in production automobiles.

Digital Gauge—See *Gauge.*

Disconnect Terminal—Solderless connectors in male and female forms, intended to be easily disconnected and connected. Typically, a blade or pin (male connector) fits into a matching receptacle or socket (female connector). Many components have built-in male (blade) terminals that require a specialized female connector.

Display—Any device that conveys information. In a car, displays are either lamps, gauges or buzzers. Gauges may be analog or digital.

DPDT—Abbreviation for double-pole, double-throw switch. See page 30.

DPST—Abbreviation for double-pole, single-throw switch. See page 30.

Electricity—A fundamental entity of nature, consisting of positive and negative charges, usually utilized in the form of currents.

Electromagnetism—Magnetism produced by electricity. The property of any conductor to be surrounded by a magnetic field whenever current is flowing. By wrapping a number of loops into a coil around a central core, the magnetic field can be concentrated for useful work.

Electron—An atomic particle that moves in an orbit around the atom's nucleus. Loosely held electrons are free to hop from one atom to another. When free electrons flow through a conductor, electrical current is produced.

Ergonomics—The science of how anatomy, physiology, psychology and other human factors relate to the design and positioning of machines, instruments and controls.

Flasher—The device that causes a turn signal or hazard warning system to turn on and off at regular intervals. Although turn and hazard flashers look alike, they are mechanically different and should not be interchanged. Hazard flashers are sometimes called *heavy duty flashers*.

Fuse—A circuit-protection device, calibrated to melt at a predetermined amp level. It acts as the "weakest link" in the circuit, shutting off current before it can damage wires or components.

Fuse Panel (Fuse Block)—A plastic or fiberboard assembly that permits mounting several fuses in one centralized location. Some fuse panels are part of, or contain, a *terminal block*.

Fusible Link—A circuit-protection device, consisting of a conductor surrounded by heat-resistant insulation. The conductor is two gages smaller than the wire it protects, so it acts as the "weakest link in the circuit." Unlike a blown fuse, a failed fusible link must be cut from the wire for replacement.

Gage—A standard SAE designation of wire sizes, expressed in AWG (American Wire Gage). The larger the gage number, the smaller the wire. Metric wire sizes are expressed in cross-sectional area, expressed in square millimeters. Also spelled *gauge:* In this book, the spelling gage is used for wire size to help avoid confusion between this and instrument-panel displays, spelled gauge (below).

Gauge—An instrument-panel display used to monitor engine conditions. Also spelled *gage* (above). A gauge with a movable pointer on a dial or fixed scale is called an *analog gauge*. Those using only numerical readouts are called *digital gauges*. The driver must interpret the gauge reading to determine what action to take, if any.

Generator—The engine-driven device that converts rotary motion into electrical potential. See also *AC Generator* and *DC Generator*.

Grommet—Component used to protect wire that passes through a panel. Usually donut shaped, made of rubber or plastic.

Ground—In a single-wire system, any metal part of the car's structure that is directly or indirectly attached to the battery negative post. Used to conduct current from a load back to the battery. *Self-grounded* components are attached directly to a grounded metal part through their mounting screws. Components mounted to nongrounded parts of a car require a separate wire running to a known good ground.

Halogen Lamp—A specialty lamp that gives off a brilliant white light. Because of its high intensity, a halogen lamp is often used for fog lamps and driving lamps.

Harness—A bundle of electrical wires. For convenience in handling and for neatness, all wires going to an area of a car are bundled into a harness.

Harness Ties—Self-tightening nylon straps used to bundle wires into harnesses. Available in stock lengths that can be cut to size after installation. Once tightened, they cannot be removed unless cut from the harness.

Harness Wrap—One of several materials used to bundle wires into manageable harnesses. See *loom, split loom, loom tape* and *harness ties*.

Human Factors—The factors of anatomy, physiology, psychology and so on, that must be considered when designing and positioning machines, instruments and controls. See *Ergonomics*.

Hydrometer—A syringelike device used to measure a battery's specific gravity.

Indicator Lamp—An instrument-panel display used to convey information or condition of the monitored circuit or system. Unlike warning lamps, indicator lamps are not used in critical situations. A headlamp high-beam indicator lamp is an example. See *Warning Lamp*.

Induction—The process of inducing voltage potential into a conductor by passing the conductor through a magnetic field. This is what causes a generator to produce electric current.

Insulator—Materials that do not conduct electrical current. Examples are cloth, glass, plastic and rubber. Wires for modern cars have plastic insulation.

Junction—Any point from which three or more wires branch out in a circuit.

Lamp—An electrical load designed to emit light when current flows through it. It consists of a glass bulb enclosing a filament, and a base containing the electrical contacts. Some lamps, such as sealed-beam headlamps, also contain a built-in reflector.

Load, Electrical—Any device that uses electrical current to perform beneficial work in a car's electrical system. Lamps and motors are the two common types of loads found in a car.

Loom—Harness covering, usually of a woven material. Older cars used woven-cloth loom, most modern cars use a corrugated-plastic loom, or *split loom*.

Loom Tape—A nonadhesive tape used as a harness wrap. Adhesive-type tapes, including electrical tape, are not recommended for wrapping harnesses. Often, a piece of shrink wrap is used at tape ends to keep tape from unraveling.

Nylon Ties—See *Harness Ties*.

Original Equipment Manufacturer (OEM)—A designation used to describe the equipment and parts installed in a car by the manufacturer, or those available from the car manufacturer as replacement parts. See *Aftermarket Parts*.

Ohm—A measure of electrical resistance.

Ohmmeter—A test instrument used to determine resistance in an electrical circuit. In auto electrical work, it is often used to determine the resistance various loads contribute to a circuit or system.

Ohm's Law—The electrical formula that describes how voltage, current and resistance are related. The basic formula is: E (electrical pressure in volts) = I (current flow in amperes) X R (resistance in ohms).

Parallel Circuit—A circuit in which current can flow independently through several components at the same time. See *Series Circuit*.

Polarity-Protected Connector—A multiple-cavity connector that can be connected in only one way, either to a mating connector or to a component.

Printed Circuit—An electrical conductor consisting of thin metal foil paths adhered to a flexible plastic backing, also called a PC board. PCs are used primarily in OEM instrument clusters and other electronic devices.

Relay—An electromechanical device consisting of an electromagnetic coil, a fixed core and a movable armature. The armature is generally connected to an electrical switch that moves according to whether the coil is energized. Like a switch, a relay is used to open and close circuits, but is usually activated by another switch in a remote location.

Resistance—In electricity, the resistance to electron flow present in an electrical circuit, expressed in ohms.

Resistor—Any conductor that permits electron movement but retards it. Examples include tungsten and nickel.

Rheostat—A variable resistor, operated by knob or handle, used to vary the resistance in a circuit. It consists of a coil of resistance wire and a movable contact or wiper that creates more or less resistance in the circuit, depending on how many coil windings it allows the current to pass through. When used as a switch, it allows a continuous setting. An example is the dimmer control for instrument-panel illumination.

Ring Terminal—A connector used to attach a wire to a screw or stud terminal. The ring is sized to the mating screw. Ring terminals are the least likely to vibrate loose in rugged applications such as racing. Comes in soldered and solderless versions.

Schematic—Another word for a wiring diagram. A drawing system for portraying the components and wires in a car's electrical system using standardized symbols.

Sending Switch—Used to operate a warning or indicator lamp. Usually a normally open SPST switch that is calibrated to close when conditions in a system reach a predetermined point or level.

Sending Unit—Used to operate a gauge. Contains a variable resistance that modifies current flow according to the condition or system being monitored. The variations in current flow are then interpreted by the gauge.

Series Circuit—A circuit in which current has to pass through a series of components, one at a time.

Series/Parallel Circuit—A circuit where some components are wired in series, others in parallel. An example is two loads wired in parallel with each other, but in series with the switch that controls them.

Short Circuit—An undesirable condition in which current is permitted to bypass the load, returning to the source without doing any useful work. "Shorts" are usually a result of a bare conductor wire accidentally touching a grounded metal part between the load and source.

Shrink Wrap—An insulating material used to protect wire splices and junctions at terminals. Upon application of open flame or heat, the wrap shrinks to fit tightly on the wire or terminal.

Simple Circuit—The simplest circuit includes an electrical source, a load and some wire to join them.

Single-Wire Circuit—Generally used in production automobiles, in which one wire brings current to the load and the car's frame acts as the return path (ground).

Snap-Splice Connector—Solderless connector used to tap an additional wire into an existing wire without cutting the original. Often used in installing trailer wiring to a tow vehicle.

Solderless Connector—Any connector or terminal that can be installed to a wire without the use of solder. They usually are crimped in place using a special crimping tool. Examples are ring terminals, spade terminals, disconnect terminals, butt connectors, closed-end connectors and snap-splice connectors. Ring and spade terminals also come in soldered versions.

Solenoid—An electromechanical device consisting of an electromagnetic coil surrounding a movable metal core, or plunger. The movable core is connected to various mechanisms to accomplish work.

Spade Terminal—A terminal used to connect a wire to a screw or stud terminal. It has two forked ends, either straight or with upturned tips. They're more convenient to install than ring terminals, but slightly less secure for rugged duty. Comes in soldered and solderless versions.

SPDT—Abbreviation for single-pole, double-throw switch. See page 30.

Specific Gravity—The measure of a battery's charge, made by comparing the weight of the electrolyte to pure water. A fully charged battery will have a specific gravity reading of 1.260. See *Hydrometer*.

Split Loom—Flexible, corrugated conduit used to bundle wires into harnesses.

SPST—Abbreviation for single-pole, single-throw switch. See page 30.

Switch—An electrical control used to turn a circuit on and off, or opening and closing a circuit. Switches are described by the number of poles and throws they have. See SPST, SPDT, DPST, AND DPDT.

Terminal Block—A plastic or resin assembly containing two rows of terminal screws. Used to join the circuits in several wiring harnesses. Occasionally contain, or are part of, a fuse panel.

Test Lamp, 12-Volt—A test instrument consisting of an indicator lamp wired into the handle of a metal probe. When the probe contacts a live circuit, current flows through the lamp, lighting it, and to ground through an attached wire and alligator clip. Used to test for voltage in live circuits only. Also called a *continuity tester*.

Test Lamp, Self Powered—A test device containing an indicator lamp and a built-in battery. Used to test continuity of circuits not containing voltage at the time of the test (dead circuits). This is the type used to test continuity in a harness before it is installed in the car.

Thermistor—The electrical element in a temperature-sending unit that varies its resistance in proportion to temperature. Unlike most electrical conductors, in which resistance increases as temperature rises, resistance in a thermistor decreases. Thermistors are made from the oxides of cobalt, copper, iron or nickel.

Tracer—A stripe of a second color applied to a wire insulator to distinguish that wire from another one with the same color insulator. Can be simulated by running a waterproof magic marker along a wire.

Volt—A measure of electrical force, or pressure.

Voltage Drop—Any condition in a circuit that causes voltage to be reduced. This could be an intentional load or an unintentional drop caused by a faulty or corroded terminal. All wire, no matter how low the resistance, shows at least a trace of voltage drop.

Voltage Regulator—An electromechanical or electronic device that monitors voltage in the vehicle's electrical system to control generator output within a narrow range. This provides enough current to operate electrical devices while preventing battery overcharge.

Voltmeter—1. An instrument-panel gauge that measures system voltage. When the engine is not running, the voltmeter indicates battery voltage, which should be 12—13 volts in a 12-volt system. When engine is running, voltmeter indicates total system voltage, or combined voltage output of generator and battery. 2. A test instrument that measures voltage in an electrical circuit. Used to check continuity and determine voltage drop in specific circuits of car's electrical system.

VOM (Volt-Ohmmeter)—A two-in-one test instrument. For convenience, a voltmeter and an ohmmeter are mounted in the same case and share a common readout and test leads.

Warning Lamp—An instrument-panel display used to inform the driver when something undesirable has happened in the monitored circuit or system, such as an overheated engine or a sudden loss of oil pressure. The driver is required to take action when a warning lamp lights. The action is often to shut down the engine to prevent damage.

Index

HP AUTOMOTIVE BOOKS

HANDBOOK SERIES

Auto Dictionary
Auto Electrical Handbook
Auto Math Handbook
Auto Paint Handbook
Baja Bugs & Buggies
Brake Handbook
Camaro Restoration Handbook
Clutch & Flywheel Handbook
Metal Fabricator's Handbook
Mustang Restoration Handbook
Off-Roader's Handbook
Paint & Body Handbook
Sheet Metal Handbook
Small Trucks
Street Rodder's Handbook
Turbochargers
Turbo Hydra-Matic 350
Welder's Handbook

CARBURETORS

Holley 4150
Holley Carburetors, Manifolds & Fuel Injection
Rochester Carburetors
Weber Carburetors

PERFORMANCE SERIES

How to Hot Rod Big-Block Chevys
How to Hot Rod Small-Block Chevys
How to Hot Rod Small-Block Mopar Engines
How to Hot Rod VW Engines
How to Make Your Car Handle
Race Car Engineering
Small-Block Chevy Performance
Chassis Engineering
Camaro Performance

REBUILD SERIES

How to Rebuild Air-Cooled VW Engines
How to Rebuild Big-Block Chevys
How to Rebuild Big-Block Ford Engines
How to Rebuild Small-Block Chevys
How to Rebuild Small-Block Ford Engines
How to Rebuild Small-Block Mopars
How to Rebuild Your Ford V-8

SPECIAL INTEREST

Auto Repair Shams & Scams
Car Collector's Handbook
Fast Fords
Guide to GM Muscle Cars

4261-22
5-43

Books are available from your local auto store, bookstore or order direct from publisher, Price Stern Sloan,
11150 Olympic Boulevard, Los Angeles, CA 90064. Call toll-free: 800/421-0892.